IARC MONOGRAPHS
ON THE
EVALUATION OF THE CARCINOGENIC RISK OF CHEMICALS TO HUMANS

Polychlorinated biphenyls
and
Polybrominated biphenyls

VOLUME 18

This publication represents the views and expert opinions
of two IARC Working Groups on the
Evaluation of the Carcinogenic Risk of Chemicals to Humans
which met in Lyon,
10-15 October 1977
and
6-13 June 1978

October 1978

INTERNATIONAL AGENCY FOR RESEARCH ON CANCER

IARC MONOGRAPHS

In 1971, the International Agency for Research on Cancer (IARC) initiated a programme on the evaluation of the carcinogenic risk of chemicals to humans involving the production of critically evaluated monographs on individual chemicals.

The objective of the programme is to elaborate and publish in the form of monographs critical reviews of data on carcinogenicity for groups of chemicals to which humans are known to be exposed, to evaluate these data in terms of human risk with the help of international working groups of experts in chemical carcinogenesis and related fields, and to indicate where additional research efforts are needed.

International Agency for Research on Cancer 1978

ISBN 92 832 1218 5

PRINTED IN SWITZERLAND

CONTENTS

LIST OF PARTICIPANTS
    PCBs 5
    PBBs 9

NOTE TO THE READER 13

PREAMBLE 15
    BACKGROUND 15
    OBJECTIVE AND SCOPE 15
    SELECTION OF CHEMICALS FOR MONOGRAPHS 16
    WORKING PROCEDURES 16
    DATA FOR EVALUATIONS 17
    THE WORKING GROUP 17
    GENERAL PRINCIPLES FOR EVALUATING THE CARCINOGENIC
    RISK OF CHEMICALS 18
    EXPLANATORY NOTES ON THE MONOGRAPH CONTENTS 23

GENERAL REMARKS ON THE SUBSTANCES CONSIDERED 29

THE MONOGRAPHS
    POLYCHLORINATED BIPHENYLS 43
        Chemical and Physical Data 43
        Production, Use, Occurrence and Analysis 54
        Biological Data Relevant to the Evaluation of Carcinogenic
            Risk to Humans 66
        Summary of Data Reported and Evaluation 83
        References 85

    POLYBROMINATED BIPHENYLS 107
        Chemical and Physical Data 107
        Production, Use, Occurrence and Analysis 110
        Biological Data Relevant to the Evaluation of Carcinogenic
            Risk to Humans 114
        Summary of Data Reported and Evaluation 119
        References 120

SUPPLEMENTARY CORRIGENDA TO VOLUMES 1-17 125

CUMULATIVE INDEX TO MONOGRAPHS 127

IARC WORKING GROUP ON THE EVALUATION OF THE CARCINOGENIC
RISK OF CHEMICALS TO HUMANS:

POLYCHLORINATED BIPHENYLS

Lyon, 10-15 October 1977

Members[1]

J.R. Allen, Professor of Pathology, University of Wisconsin, The Medical School Department of Pathology, 470 North Charter Street, Madison, Wisconsin 53706, United States of America

B.K. Armstrong, University of Western Australia, Department of Medicine, Medical School Building, Queen Elizabeth II Medical Centre, Nedlands, Western Australia 6009, Australia

P. Bannasch, Abteilung für Cytopathologie, Institut für Experimentelle Pathologie, Deutsches Krebsforschungszentrum, Postfach, 6900 Heidelberg 1, Federal Republic of Germany

G. Bochert, Institut für Toxikologie und Embryonal-Pharmakologie der Freien Universität Berlin, Garystrasse 9, 1000 Berlin 33, Federal Republic of Germany

D.H. Fine, Senior Scientist, Thermo Electron Research Center, 101 First Avenue, Waltham, Massachusetts 02154, United States of America

W. Lijinsky, Director, Chemical Carcinogenesis Program, Frederick Cancer Research Center, PO Box B, Frederick, Maryland 21701, United States of America

P.N. Magee, Director, Fels Research Institute, Temple University, School of Medicine, Philadelphia, Pennsylvania 19140, United States of America (*Chairman*)

U. Mohr, Director, Abteilung für Experimentelle Pathologie, Medizinische Hochschule Hannover, Karl-Wiechert-Allee 9, 3000 Hannover 61, Federal Republic of Germany

A.E. Pegg, Professor, Department of Physiology, The Milton S. Hershey Medical Center, The Pennsylvania State University, Hershey, Pennsylvania 17033, United States of America

---

[1]Unable to attend: G. Eisenbrand, Institut für Toxikologie und Chemotherapie, Deutsches Krebsforschungszentrum, Im Neuenheimer Feld 280, 6900 Heidelberg 1, Federal Republic of Germany

R. Preussmann, Institut für Toxikologie und Chemotherapie, Deutsches
   Krebsforschungszentrum, Im Neuenheimer Feld 280, 6900 Heidelberg 1,
   Federal Republic of Germany (*Vice-Chairman*)

C. Rappe, Department of Organic Chemistry, Umeå University, S-901 87 Umeå,
   Sweden

B.W. Stewart, School of Pathology, University of New South Wales,
   PO Box 1, Kensington, New South Wales 2033, Australia

S.R. Tannenbaum, Professor of Food Chemistry, Department of Nutrition and
   Food Science, Massachusetts Institute of Technology, Cambridge,
   Massachusetts 02139, United States of America

E. Vogel, Department of Radiation Genetics and Chemical Mutagenesis of
   the State University of Leiden, Wassenaarseweg 72, Leiden, The
   Netherlands

Representative from the National Cancer Institute

S. Siegel, Coordinator, Technical Information Activities, Technical
   Information Resources Branch, Room 3A-06, Landow Building,
   Carcinogenesis Bioassay Testing Program, Division of Cancer Cause
   and Prevention, National Cancer Institute, Bethesda, Maryland 20014,
   United States of America

Representative from SRI International

K.E. McCaleb, Director, Chemical-Environmental Program, Chemical
   Industries Center, SRI International, Menlo Park, California 94025,
   United States of America (*Rapporteur sections 2.1 and 2.2*)

Representative from the Commission of the European Communities

M.-T. van der Venne, Commission of the European Communities, Health and
   Safety Directorate, Bâtiment Jean Monnet, Plateau du Kirchberg,
   Boîte Postale 1907, Luxembourg, Great Duchy of Luxembourg

Representative from the United Nations Environment Programme

H.E. Christensen, Chief, Information Processing Unit, Room 31,
   International Register of Potentially Toxic Chemicals, United Nations
   Environment Programme, World Health Organization, 1211 Geneva 27,
   Switzerland

Secretariat

- H. Bartsch, Unit of Chemical Carcinogenesis (*Rapporteur section 3.2*)
- J. Cooper, Unit of Epidemiology and Biostatistics (*Co-rapporteur section 3.3*)
- L. Griciute, Chief, Unit of Environmental Carcinogens
- J.E. Huff, Unit of Chemical Carcinogenesis (*Secretary*)
- D. Mietton, Unit of Chemical Carcinogenesis (*Library assistant*)
- Dr R. Montesano, Unit of Chemical Carcinogenesis (*Rapporteur section 3.1*)
- C. Partensky, Unit of Chemical Carcinogenesis (*Technical editor*)
- I. Peterschmitt, Unit of Chemical Carcinogenesis, WHO, Geneva (*Bibliographic researcher*)
- V. Ponomarkov, Unit of Chemical Carcinogenesis
- R. Saracci, Unit of Epidemiology and Biostatistics (*Rapporteur section 3.3*)
- L. Tomatis, Chief, Unit of Chemical Carcinogenesis (*Head of the Programme*)
- E.A. Walker, Unit of Environmental Carcinogens (*Rapporteur sections 1 and 2.3*)
- E. Ward, Montignac, France (*Editor*)
- J.D. Wilbourn, Unit of Chemical Carcinogenesis (*Co-secretary*)

Secretarial assistance

- A.V. Anderson
- M.-J. Ghess
- R.B. Johnson

IARC WORKING GROUP ON THE EVALUATION OF THE CARCINOGENIC
RISK OF CHEMICALS TO HUMANS:

POLYBROMINATED BIPHENYLS

Lyon, 6-13 June 1978

Members

O. Axelson, Department of Occupational Medicine, University Hospital,
   S-581 85 Linköping, Sweden

J.R.P. Cabral, MRC Toxicology Unit, Medical Research Council Laboratories,
   Woodmansterne Road, Carshalton, Surrey SM5 4EF, United Kingdom

I. Chernozemsky, Chief, Laboratory of Carcinogenesis, Institute of
   Oncology, Medical Academy, Sofia 1156, Bulgaria

C. Cueto, Jr, Chief, Toxicology Branch, Carcinogenesis Testing Program,
   Division of Cancer Cause and Prevention, National Cancer Institute,
   Bethesda, Maryland 20014, United States of America

S.S. Epstein, Professor of Occupational and Environmental Medicine,
   School of Public Health, University of Illinois at the Medical
   Center, Chicago, PO Box 6998, Chicago, Illinois 60680, United States
   of America

R. Gingell, Assistant Professor, The Eppley Institute for Research in
   Cancer, University of Nebraska Medical Center, 42nd and Dewey Avenue,
   Omaha, Nebraska 68105, United States of America

K.S. Larsson, Laboratory of Teratology, Karolinska Institute, S-104 01
   Stockholm, Sweden

J.A. Moore, Associate Director, Research Resources Program, National
   Institute of Environmental Health Sciences, PO Box 12233, Research
   Triangle Park, North Carolina 27709, United States of America

N. Nelson, Professor and Chairman, Institute of Environmental Medicine,
   New York University Medical Center, 550 First Avenue, New York,
   New York 10016, United States of America (*Chairman*)

K. Sankaranarayanan, Associate Professor, Department of Radiation Genetics
   and Chemical Mutagenesis, Sylvius Laboratories, State University of
   Leiden, Wassenaarseweg 72, Leiden, The Netherlands

B. Teichmann, Department of Chemical Carcinogenesis, Zentralinstitut
   für Krebsforschung, Akademie der Wissenschaften der DDR, Lindenberger
   Weg 80, 115 Berlin-Buch, German Democratic Republic

B. Terracini, Epidemiologia dei Tumori, Istituto di Anatomia e Istologia
Patologica dell'Università di Torino, Via Santena 7, 10126 Torino,
Italy (*Vice-Chairman*)

R. Truhaut, Director, Centre de Recherches Toxicologiques, Faculté des
Sciences pharmaceutiques et biologiques de l'Université René Descartes,
4 Avenue de l'Observatoire, 75006 Paris, France

H. Uehleke, Chief Director, Toxicology, Department of Toxicology,
Bundesgesundheitsamt, Thielallee 88/92, 1 Berlin 33-Postfach, Federal
Republic of Germany

S. Venitt, Institute of Cancer Research, Pollards Wood Research Station,
Nightingales Lane, Chalfont St Giles, Bucks HP8 4SP, United Kingdom

J.K. Wagoner, Special Assistant for Occupational Carcinogenesis, Office
of the Assistant Secretary of Labor, US Occupational Safety and Health
Administration, US Department of Labor, 200 Constitution Avenue NW,
Washington DC 20210, United States of America

J.S. Wassom, Director, Environmental Mutagen and Teratology Information
Centers, Oak Ridge National Laboratory, PO Box Y, Oak Ridge,
Tennessee 37830, United States of America

F. Wiebel, Institute of Toxicology and Biochemistry, Gesellschaft für
Strahlen- und Umweltforschung MBH München, Ingolstädter Landstrasse 1,
Post Oberschleissheim, 8042 Neuherberg, Federal Republic of Germany

Observers

A.M. Kaplan, Chief, Oral Toxicology, Haskell Laboratory for Toxicology
and Industrial Medicine, E.I. Du Pont de Nemours & Co., Inc.,
Elkton Road, Newark, Delaware 19711, United States of America

L. Villemey, Délégué à l'Environnement, European Council of Chemical
Manufacturers' Federations, Avenue Louise 250, 1050 Brussels, Belgium

Representative from SRI International

O.H. Johnson, Senior Industrial Economist, Chemical-Environmental Program,
SRI International, 333 Ravenswood Avenue, Menlo Park, California 94025,
United States of America (*Rapporteur sections 2.1 and 2.2*)

Representative from the Commission of the European Communities

M.-T. van der Venne, Commission of the European Communities, Health and
Safety Directorate, Bâtiment Jean Monnet, Plateau du Kirchberg,
Bôite Postale 1907, Luxembourg, Great Duchy of Luxembourg

Secretariat

- C. Agthe, Chief Scientist, Responsible Officer (Food Safety), Environmental Health Criteria and Standards, Division of Environmental Health, WHO, Geneva

- H. Bartsch, Unit of Chemical Carcinogenesis (*Rapporteur section 3.2*)

- M. Castegnaro, Unit of Environmental Carcinogens

- J. Cooper, Unit of Epidemiology and Biostatistics (*Co-rapporteur section 3.3*)

- L. Griciute, Chief, Unit of Environmental Carcinogens

- J.E. Huff, Unit of Chemical Carcinogenesis (*Secretary*)

- T. Kuroki, Unit of Chemical Carcinogenesis

- D. Mietton, Unit of Chemical Carcinogenesis (*Library assistant*)

- R. Montesano, Unit of Chemical Carcinogenesis (*Rapporteur section 3.1*)

- C. Partensky, Unit of Chemical Carcinogenesis (*Technical editor*)

- I. Peterschmitt, Unit of Chemical Carcinogenesis, WHO, Geneva (*Bibliographic researcher*)

- V. Ponomarkov, Unit of Chemical Carcinogenesis

- R. Saracci, Unit of Epidemiology and Biostatistics (*Rapporteur section 3.3*)

- L. Tomatis, Chief, Unit of Chemical Carcinogenesis (*Head of the Programme*)

- E.A. Walker, Unit of Environmental Carcinogens (*Rapporteur sections 1 and 2.3*)

- E. Ward, Montignac, France (*Editor*)

- J.D. Wilbourn, Unit of Chemical Carcinogenesis (*Co-secretary*)

Secretarial assistance

- A.V. Anderson
- M.-J. Ghess
- R.B. Johnson

NOTE TO THE READER

The term 'carcinogenic risk' in the IARC Monograph series is taken to mean the probability that exposure to the chemical will lead to cancer in humans.

Inclusion of a chemical in the monographs does not imply that it is a carcinogen, only that the published data have been examined. Equally, the fact that a chemical has not yet been evaluated in a monograph does not mean that it is not carcinogenic.

Anyone who is aware of published data that may alter the evaluation of the carcinogenic risk of a chemical for humans is encouraged to make this information available to the Unit of Chemical Carcinogenesis, International Agency for Research on Cancer, Lyon, France, in order that the chemical may be considered for re-evaluation by a future Working Group.

Although every effort is made to prepare the monographs as accurately as possible, mistakes may occur. Readers are requested to communicate any errors to the Unit of Chemical Carcinogenesis, so that corrections can be reported in future volumes.

IARC MONOGRAPH PROGRAMME ON THE EVALUATION OF THE
CARCINOGENIC RISK OF CHEMICALS TO HUMANS

PREAMBLE

BACKGROUND

In 1971, the International Agency for Research on Cancer (IARC) initiated a programme on the evaluation of the carcinogenic risk of chemicals to humans with the object of producing monographs on individual chemicals*. The criteria established at that time to evaluate carcinogenic risk to humans were adopted by all the working groups whose deliberations resulted in the first 16 volumes of the *IARC Monograph* series. In October 1977, a joint IARC/WHO *ad hoc* Working Group met to re-evaluate these guiding criteria; this preamble reflects the results of their deliberations(1) and those of a subsequent IARC *ad hoc* Working Group which met in April 1978(2).

OBJECTIVE AND SCOPE

The objective of the programme is to elaborate and publish in the form of monographs critical reviews of data on carcinogenicity for groups of chemicals to which humans are known to be exposed, to evaluate these data in terms of human risk with the help of international working groups of experts in chemical carcinogenesis and related fields, and to indicate where additional research efforts are needed.

The monographs summarize the evidence for the carcinogenicity of individual chemicals and other relevant information. The critical analyses of the data are intended to assist national and international authorities in formulating decisions concerning preventive measures. No recommendations are given concerning legislation, since this depends on risk-benefit evaluations, which seem best made by individual governments and/or international agencies. In this connection, WHO recommendations on food additives(3), drugs(4), pesticides and contaminants(5) and occupational carcinogens(6) are particularly informative.

---

*Since 1972, the programme has undergone considerable expansion, primarily with the scientific collaboration and financial support of the US National Cancer Institute.

The *IARC Monographs* are recognized as an authoritative source of information on the carcinogenicity of environmental chemicals. The first users' survey, made in 1976, indicates that the monographs are consulted routinely by various agencies in 24 countries.

Since the programme began in 1971, 18 volumes have been published(7-24) in the *IARC Monograph* series, and 381 separate chemical substances have been evaluated (see also cumulative index to the monographs, p. 127). Each volume is printed in 4000 copies and distributed *via* the WHO publications service (see inside covers for a listing of IARC publications and back outside cover for distribution and sales services).

SELECTION OF CHEMICALS FOR MONOGRAPHS

The chemicals (natural and synthetic, including those which occur as mixtures and in manufacturing processes) are selected for evaluation on the basis of two main criteria: (a) there is evidence of human exposure, and (b) there is some experimental evidence of carcinogenicity and/or there is some evidence or suspicion of a risk to humans. In certain instances, chemical analogues were also considered.

Inclusion of a chemical in a volume does not imply that it is carcinogenic, only that the published data have been examined. The evaluations must be consulted to ascertain the conclusions of the Working Group. Equally, the fact that a chemical has not appeared in a monograph does not mean that it is without carcinogenic hazard.

The scientific literature is surveyed for published data relevant to the monograph programme. In addition, the IARC Survey of Chemicals Being Tested for Carcinogenicity(25-31) often indicates those chemicals that are to be scheduled for future meetings. The major aims of the survey are to prevent unnecessary duplication of research, to increase communication among scientists, and to make a census of chemicals that are being tested and of available research facilities.

As new data on chemicals for which monographs have already been prepared and new principles for evaluating carcinogenic risk receive acceptance, re-evaluations will be made at subsequent meetings, and revised monographs will be published as necessary.

WORKING PROCEDURES

Approximately one year in advance of a meeting of a working group, a list of the substances to be considered is prepared by IARC staff in consultation with other experts. Subsequently, all relevant biological data are collected by IARC; in addition to the published literature, US Public Health Service Publication No. 149(32-37) has been particularly valuable and has been used in conjunction with other recognized sources of information on chemical carcinogenesis and systems such as MEDLINE and TOXLINE. The major collection of data and the preparation of first drafts for the sections on chemical and physical properties, on production, use, occurrence and on analysis are carried out by Stanford Research Institute (SRI) International under a separate contract with the US National Cancer Institute. Most of the data so obtained on production, use and occurrence refer to the United States and Japan; SRI and IARC supplement this information with that from other sources in Europe. Bibliographical sources for data on mutagenicity and teratogenicity are the Environmental Mutagen Information Center and the Environmental Teratology Information Center, both located at the Oak Ridge National Laboratory, USA.

Six to nine months before the meeting, reprints of articles containing relevant biological data are sent to an expert(s), or are used by the IARC staff, for the preparation of first drafts of the monographs. These drafts are edited by IARC staff and are sent prior to the meeting to all participants of the Working Group for their comments. The Working Group then meets in Lyon for seven to eight days to discuss and finalize the texts of the monographs and to formulate the evaluations. After the meeting, the master copy of each monograph is verified by consulting the original literature, then edited and prepared for reproduction. The monographs are usually published within six months after the Working Group meeting.

DATA FOR EVALUATIONS

With regard to biological data, only reports that have been published or accepted for publication are reviewed by the working groups, although a few exceptions have been made. The monographs do not cite all of the literature on a particular chemical: only those data considered by the Working Group to be relevant to the evaluation of the carcinogenic risk of the chemical to humans are included.

Anyone who is aware of data that have been published or are in press which are relevant to the evaluations of the carcinogenic risk to humans of chemicals for which monographs have appeared is urged to make them available to the Unit of Chemical Carcinogenesis, International Agency for Research on Cancer, Lyon, France.

THE WORKING GROUP

The tasks of the Working Group are five-fold: (a) to ascertain that all data have been collected; (b) to select the data relevant for the evaluation; (c) to ensure that the summaries of the data enable the reader to follow the reasoning of the committee; (d) to judge the significance of the results of experimental and epidemiological studies; and (e) to make an evaluation of the carcinogenic risk of the chemical.

Working Group participants who contributed to the consideration and evaluation of chemicals within a particular volume are listed, with their addresses, at the beginning of each publication (see p. 3). Each member serves as an individual scientist and not as a representative of any organization or government. In addition, observers are often invited from national and international agencies, organizations and industries.

GENERAL PRINCIPLES FOR EVALUATING THE CARCINOGENIC RISK OF CHEMICALS

The widely accepted meaning of the term 'chemical carcinogenesis', and that used in these monographs, is the induction by chemicals of neoplasms that are not usually observed, the earlier induction by chemicals of neoplasms that are usually observed, and/or the induction by chemicals of more neoplasms than are usually found - although fundamentally different mechanisms may be involved in these three situations. Etymologically, the term 'carcinogenesis' means the induction of cancer, that is, of malignant neoplasms; however, the commonly accepted meaning is the induction of various types of neoplasms or of a combination of malignant and benign tumours. In the monographs, the words 'tumour' and neoplasm' are used interchangeably (In scientific literature the terms 'tumourigen', 'oncogen' and 'blastomogen' have all been used synonymously with 'carcinogen', although occasionally 'tumourigen' has been used specifically to denote the induction of benign tumours).

Experimental Evidence

Qualitative aspects

Both the interpretation and evaluation of a particular study as well as the overall assessment of the carcinogenic activity of a chemical involve several qualitatively important considerations, including:
(a) the experimental parameters under which the chemical was tested, including route of administration and exposure, species, strain, sex, age, etc.; (b) the consistency with which the chemical has been shown to be carcinogenic, e.g., in how many species and at which target organ(s); (c) the spectrum of neoplastic response, from benign neoplasia to multiple malignant tumours; (d) the stage of tumour formation in which a chemical may be involved: some chemicals act as complete carcinogens and have initiating and promoting activity, while others are promoters only; and (e) the possible role of modifying factors.

There are problems not only of differential survival but of differential toxicity, which may be manifested by unequal growth and weight gain in treated and control animals. These complexities should also be considered in the interpretation of data, or, better, in the experimental design.

Many chemicals induce both benign and malignant tumours; few instances are recorded in which only benign neoplasms are induced by chemicals that have been studied extensively. Benign tumours may represent a stage in the evolution of a malignant neoplasm or they may be 'end-points' that do not readily undergo transition to malignancy. If a substance is found to induce only benign tumours in experimental animals, the chemical should be suspected of being a carcinogen and requires further investigation.

Hormonal carcinogenesis

Hormonal carcinogenesis presents certain distinctive features: the chemicals involved occur both endogenously and exogenously; in many instances, long exposure is required; tumours occur in the target issue in association with a stimulation of non-neoplastic growth, but in some cases, hormones promote the proliferation of tumour cells in a target organ. Hormones that occur in excessive amounts, hormone-mimetic agents and agents that cause hyperactivity or imbalance in the endocrine system may require evaluative methods comparable with those used to identify chemical carcinogens; particular emphasis must be laid on quantitative aspects and duration of exposure. Some chemical carcinogens have significant side effects on the endocrine system, which may also result in hormonal carcinogenesis. Synthetic hormones and anti-hormones can be expected to possess other pharmacological and toxicological actions in addition to those on the endocrine system, and in this respect they must be treated like any other chemical with regard to intrinsic carcinogenic potential.

Quantitative aspects

Dose-response studies are important in the evaluation of carcinogenesis: the confidence with which a carcinogenic effect can be established is strengthened by the observation of an increasing incidence of neoplasms with increasing exposure.

The assessment of carcinogenicity in animals is frequently complicated by recognized differences among the test animals (species, strain, sex, age), in route(s) of administration and in dose/duration of exposure; often, target organs at which a cancer occurs and its histological type may vary with these parameters. Nevertheless, indices of carcinogenic potency in particular experimental systems (for instance, the dose-rate required under continuous exposure to halve the probability of the animals remaining tumourless(38)) have been formulated in the hope that, at least among categories of fairly similar agents, such indices may be of some predictive value in other systems, including humans.

Chemical carcinogens differ widely in the dose required to produce a given level of tumour induction, although many of them share common biological properties which include metabolism to reactive (electrophilic (39-41)) intermediates capable of interacting with DNA. The reason for this variation in dose-response is not understood but may be due either to differences within a common metabolic process or to the operation of qualitatively distinct mechanisms.

Statistical analysis of animal studies

Tumours which would have arisen had an animal lived longer may not be observed because of the death of the animal from unrelated causes, and this possibility must be allowed for. Various analytical techniques have been developed which use the assumption of independence of competing risks to allow for the effects of intercurrent mortality on the final numbers of tumour-bearing animals in particular treatment groups.

For externally visible tumours and for neoplasms that cause death, methods such as Kaplan-Meier (i.e., 'life-table', 'product-limit' or 'actuarial') estimates(38), with associated significance tests(42,43), are recommended.

For internal neoplasms which are discovered 'incidentally'(42) at autopsy but which did not cause the death of the host, different estimates (44) and significance tests(42,43) may be necessary for the unbiased study of the numbers of tumour-bearing animals.

All of these methods(38,42-44) can be used to analyse the numbers of animals bearing particular tumour types, but they do not distinguish between animals with one or many such tumours. In experiments which end at a particular fixed time, with the simultaneous sacrifice of many animals, analysis of the total numbers of internal neoplasms per animal found at autopsy at the end of the experiment is straightforward. However, there are no adequate statistical methods for analysing the numbers of particular neoplasms that kill an animal host.

Evidence of Carcinogenicity in Humans

Evidence of carcinogenicity in humans can be derived from three types of study, the first two of which usually provide only suggestive evidence: (a) reports concerning individual cancer patients (case reports), including a history of exposure to the supposed carcinogenic agent; (b) descriptive epidemiological studies in which the incidence of cancer in human populations is found to vary (spatially or temporally) with exposure to the agent; and (c) analytical epidemiological studies (e.g., case-control or cohort studies) in which individual exposure to the agent is found to be associated with an increased risk of cancer.

An analytical study that shows a positive association between an agent and a cancer may be interpreted as implying causality to a greater or lesser extent, if the following criteria are met: (a) there is no identifiable positive bias (By 'positive bias' is meant the operation of factors in study design or execution which lead erroneously to a more strongly positive association between an agent and disease than in fact exists. Examples of positive bias include, in case-control studies, better documentation of exposure to the agent for cases than for controls, and, in cohort studies, the use of better means of detecting cancer in individuals exposed to the agent than in individuals not exposed);

(b) the possibility of positive confounding has been considered (By 'positive confounding' is meant a situation in which the relationship between an agent and a disease is rendered more strongly positive than it truly is as a result of an association between that agent and another agent which either causes or prevents the disease. An example of positive confounding is the association between coffee consumption and lung cancer, which results from their joint association with cigarette smoking); (c) the association is unlikely to be due to chance alone; (d) the association is strong; and (e) there is a dose-response relationship.

In some instances, a single epidemiological study may be strongly indicative of a cause-effect relationship; however, the most convincing evidence of causality comes when several independent studies done under different circumstances result in 'positive' findings.

Analytical epidemiological studies that show no association between an agent and a cancer ('negative' studies) should be interpreted according to criteria analogous to those listed above: (a) there is no identifiable negative bias; (b) the possibility of negative confounding has been considered; and (c) the possible effects of misclassification of exposure or outcome have been weighed.

In addition, it must be recognized that in any study there are confidence limits around the estimate of association or relative risk. In a study regarded as 'negative', the upper confidence limit may indicate a relative risk substantially greater than unity; in that case, the study excludes only relative risks that are above this upper limit. This usually means that a 'negative' study must be large to be convincing. Confidence in a 'negative' result is increased when several independent studies carried out under different circumstances are in agreement.

Finally, a 'negative' study may be considered to be relevant only to dose levels within or below the range of those observed in the study and is pertinent only if sufficient time has elapsed since first human exposure to the agent. Experience with human cancers of known etiology suggests that the period from first exposure to a chemical carcinogen to development of clinically observed cancer is usually measured in decades and may be in excess of 30 years.

Experimental Data Relevant to the Evaluation of Carcinogenic Risk to Humans

No adequate criteria are presently available to interpret experimental carcinogenicity data directly in terms of carcinogenic potential for humans. Nonetheless, utilizing data collected from appropriate tests in animals, positive extrapolations to possible human risk can be approximated.

Information compiled from the first 17 volumes of the *IARC Monographs* (45-47) shows that of about 26 chemicals or manufacturing processes now generally accepted to cause cancer in humans, all but possibly two (arsenic and benzene) of those which have been tested appropriately produce cancer in at least one animal species. For several (aflatoxins, 4-aminobiphenyl, diethylstilboestrol, melphalan, mustard gas and vinyl chloride), evidence of carcinogenicity in experimental animals preceded evidence obtained from epidemiological studies or case reports.

In general, the evidence that a chemical produces tumours in experimental animals is of two degrees: (a) *sufficient evidence* of carcinogenicity is provided by the production of malignant tumours; and (b) *limited evidence* of carcinogenicity reflects qualitative and/or quantitative limitations of the experimental results.

For many of the chemicals evaluated in the first 17 volumes of the *IARC Monographs* for which there is *sufficient evidence* of carcinogenicity in animals, data relating to carcinogenicity for humans are either insufficient or nonexistent. In the absence of adequate data on humans, it is reasonable, for practical purposes, to regard such chemicals as if they were carcinogenic to humans.

Sufficient evidence of carcinogenicity is provided by experimental studies that show an increased incidence of malignant tumours: (i) in multiple species or strains, and/or (ii) in multiple experiments (routes and/or doses), and/or (iii) to an unusual degree (with regard to incidence, site, type and/or precocity of onset). Additional evidence may be provided by data concerning dose-response, mutagenicity or structure.

In the present state of knowledge, it would be difficult to define a predictable relationship between the dose (mg/kg bw/day) of a particular chemical required to produce cancer in test animals and the dose which would produce a similar incidence of cancer in humans. The available data suggest, however, that such a relationship may exist(48,49), at least for certain classes of carcinogenic chemicals. Data that provide *sufficient evidence* of carcinogenicity in test animals may therefore be used in an approximate quantitative evaluation of the human risk at some given exposure level, provided that the nature of the chemical concerned and the physiological, pharmacological and toxicological differences between the test animals and humans are taken into account. However, no acceptable methods are currently available for quantifying the possible errors in such a procedure, whether it is used to generalize between species or to extrapolate from high to low doses. The methodology for such quantitative extrapolation to humans requires further development.

Evidence for the carcinogenicity of some chemicals in experimental animals may be *limited* for two reasons. Firstly, experimental data may be restricted to such a point that it is not possible to determine a causal relationship between administration of a chemical and the development of a particular lesion in the animals. Secondly, there are certain neoplasms, including lung tumours and hepatomas in mice, which have been considered of lesser significance than neoplasms occurring at other sites for the purpose of evaluating the carcinogenic risk of chemicals to humans. Such tumours occur spontaneously in high incidence in these animals, and their malignancy is often difficult to establish. An evaluation of the significance of these tumours following administration of a chemical is the responsibility of particular Working Groups preparing individual monographs, and it has not been possible to set down rigid guidelines; the relevance of these tumours must be determined by considerations which include experimental design and completeness of reporting.

Some chemicals for which there is *limited evidence* of carcinogenicity in animals have also been studied in humans with, in general, inconclusive results. While such chemicals may indeed be carcinogenic to humans, more experimental and epidemiological investigation is required.

Hence, '*sufficient evidence*' of carcinogenicity and '*limited evidence*' of carcinogenicity do not indicate categories of chemicals: the inherent definitions of those terms indicate varying degrees of experimental evidence, which may change if and when new data on the chemicals become available. The main drawback to any rigid classification of chemicals with regard to their carcinogenic capacity is the as yet incomplete knowledge of the mechanism(s) of carcinogenesis.

In recent years, several short-term tests for the detection of potential carcinogens have been developed. When only inadequate experimental data are available, positive results in validated short-term tests (see p. 26) are an indication that the compound is a potential carcinogen and that it should be tested in animals for an assessment of its carcinogenicity. Negative results from short-term tests cannot be considered sufficient evidence to rule out carcinogenicity. Whether short-term tests will eventually be as reliable as long-term tests in predicting carcinogenicity in humans will depend on further demonstrations of consistency with long-term experiments and with data from humans.

EXPLANATORY NOTES ON THE MONOGRAPH CONTENTS

Chemical and Physical Data (Section 1)

The Chemical Abstracts Service Registry Number and the latest Chemical Abstracts Primary Name (9th Collective Index)(50) are recorded in section 1. Other synonyms and trade names are given, but no comprehensive list is provided. Further, some of the trade names are those of mixtures in which the compound being evaluated is only one of the ingredients.

The structural and molecular formulae, molecular weight and chemical and physical properties are given. The properties listed refer to the pure substance, unless otherwise specified, and include, in particular, data that might be relevant to carcinogenicity (e.g., lipid solubility) and those that concern identification. A separate description of the composition of technical products includes available information on impurities and formulated products.

Production, Use, Occurrence and Analysis (Section 2)

The purpose of section 2 is to provide indications of the extent of past and present human exposure to this chemical.

Synthesis

Since cancer is a delayed toxic effect, the dates of first synthesis and of first commercial production of the chemical are provided. In addition, methods of synthesis used in past and present commercial production are described. This information allows a reasonable estimate to be made of the date before which no human exposure could have occurred.

Production

Since Europe, Japan and the United States are reasonably representative industrialized areas of the world, most data on production, foreign trade and uses are obtained from those countries. It should not, however, be inferred that those nations are the sole or even the major sources or users of any individual chemical.

Production and foreign trade data are obtained from both governmental and trade publications by chemical economists in the three geographical areas. In some cases, separate production data on organic chemicals manufactured in the United States are not available because their publication could disclose confidential information. In such cases, an indication of the minimum quantity produced can be inferred from the number of companies reporting commercial production. Each company is required to report on individual chemicals if the sales value or the weight of the annual production exceeds a specified minimum level. These levels vary for chemicals classified for different uses, e.g., medicinals and plastics; in fact, the minimal annual sales value is between $1000 and $50 000, and the minimal annual weight of production is between 450 and 22 700 kg. Data on production in some European countries are obtained by means of general questionnaires sent to companies thought to produce the compounds being evaluated. Information from the completed questionnaires is compiled by country, and the resulting estimates of production are included in the individual monographs.

Use

Information on uses is meant to serve as a guide only and is not complete. It is usually obtained from published data but is often complemented by direct contact with manufacturers of the chemical. In the case of drugs, mention of their therapeutic uses does not necessarily represent current practice nor does it imply judgement as to their clinical efficacy.

Statements concerning regulations and standards (e.g., pesticide registrations, maximum levels permitted in foods, occupational standards and allowable limits) in specific countries are mentioned as examples only. They may not reflect the most recent situation, since such legislation is

PREAMBLE

in a constant state of change; nor should it be taken to imply that other countries do not have similar regulations.

Occurrence

Information on the occurrence of a chemical in the environment is obtained from published data, including that derived from the monitoring and surveillance of levels of the chemical in occupational environments, air, water, soil, foods and tissues of animals and humans. When available, data on the generation, persistence and bioaccumulation of a chemical are also included.

Analysis

The purpose of the section on analysis is to give the reader an indication, rather than a complete review, of methods cited in the literature. No attempt is made to evaluate critically or to recommend any of the methods.

Biological Data Relevant to the Evaluation of Carcinogenic Risk to Humans (Section 3)

In general, the data recorded in section 3 are summarized as given by the author; however, comments made by the Working Group on certain shortcomings of reporting, of statistical analysis or of experimental design are given in square brackets. The nature and extent of impurities/contaminants in the chemicals being tested are given when available.

Carcinogenicity and related studies in animals

The monographs are not intended to cover all reported studies. Some studies are purposely omitted (a) because they are inadequate, as judged from previously described criteria(51-54) (e.g., too short a duration, too few animals, poor survival); (b) because they only confirm findings that have already been fully described; or (c) because they are judged irrelevant for the purpose of the evaluation. In certain cases, however, such studies are mentioned briefly, particularly when the information is considered to be a useful supplement to other reports or when it is the only data available. Their inclusion does not, however, imply acceptance of the adequacy of their experimental design and/or of the analysis and interpretation of their results.

Mention is made of all routes of administration by which the compound has been adequately tested and of all species in which relevant tests have been done(5,54). In most cases, animal strains are given (General characteristics of mouse strains have been reviewed(55)). Quantitative data are given to indicate the order of magnitude of the effective carcinogenic doses. In general, the doses and schedules are indicated as they appear in the original paper; sometimes units have been converted for easier comparison. Experiments on the carcinogenicity of known metabolites, chemical precursors, analogues and derivatives, and experiments on factors that modify the carcinogenic effect are also reported.

Other relevant biological data

Lethality data are given when available, and other data on toxicity are included when considered relevant. The metabolic data are restricted to studies that show the metabolic fate of the chemical in animals and humans, and comparisons of data from animals and humans are made when possible. Information is also given on absorption, distribution, excretion and placental transfer.

Embryotoxicity and teratogenicity

Data on teratogenicity from studies in experimental animals and from observations in humans are also included. There appears to be no causal relationship between teratogenicity(56) and carcinogenicity, but chemicals often have both properties. Evidence of teratogenicity suggests transplacental transfer, which is a prerequisite for transplacental carcinogenesis.

Indirect tests (mutagenicity and other short-term tests)

Data from indirect tests are also included. Since most of these tests have the advantage of taking less time and being less expensive than mammalian carcinogenicity studies, they are generally known as 'short-term' tests. They comprise assay procedures which rely on the induction of biological and biochemical effects in *in vivo* and/or *in vitro* systems. The end-point of the majority of these tests is the production not of neoplasms in animals but of changes at the molecular, cellular or multi-cellular level: these include the induction of DNA damage and repair, mutagenesis in bacteria and other organisms, transformation of mammalian cells in culture, and other systems.

The short-term tests are proposed for use (a) in predicting potential carcinogenicity in the absence of carcinogenicity data in animals, (b) as a contribution in deciding which chemicals should be tested in animals, (c) in identifying active fractions of complex mixtures containing carcinogens, (d) for recognizing active metabolites of known carcinogens in human and/or animal body fluids and (e) to help elucidate mechanisms of carcinogenesis.

Although the theory that cancer is induced as a result of somatic mutation suggests that agents which damage DNA *in vivo* may be carcinogens, the precise relevance of short-term tests to the mechanism by which cancer is induced is not known. Predictions of potential carcinogenicity are currently based on correlations between responses in short-term tests and data from animal carcinogenicity and/or human epidemiological studies. This approach is limited because the number of chemicals known to be carcinogenic in humans is insufficient to provide a basis for validation, and most validation studies involve chemicals that have been evaluated for carcinogenicity only in animals. The selection of chemicals is in turn limited to those classes for which data on carcinogenicity are available. The results of validation studies could be strongly influenced by such selection of chemicals and by the proportion of carcinogens in the series of chemicals tested; this should be kept in mind when evaluating the predictivity of a particular test. The usefulness of any test is reflected by its ability to classify carcinogens and noncarcinogens, using the animal data as a standard; however, animal tests may not always provide a perfect standard. The attainable level of correlation between short-term tests and animal bioassays is still under investigation.

Since many chemicals require metabolism to an active form, tests that do not take this into account may fail to detect certain potential carcinogens. The metabolic activation systems used in short-term tests (e.g., the cell-free systems used in bacterial tests) are meant to approximate the metabolic capacity of the whole organism. Each test has its advantages and limitations; thus, more confidence can be placed in the conclusions when negative or positive results for a chemical are confirmed in several such test systems. Deficiencies in metabolic competence may lead to misclassification of chemicals, which means that not all tests are suitable for assessing the potential carcinogenicity of all classes of compounds.

The present state of knowledge does not permit the selection of a specific test(s) as the most appropriate for identifying potential carcinogenicity. Before the results of a particular test can be considered to be fully acceptable for predicting potential carcinogenicity, certain criteria should be met: (a) the test should have been validated with respect to known animal carcinogens and found to have a high capacity for discriminating between carcinogens and noncarcinogens, and (b), when possible, a structurally related carcinogen(s) and noncarcinogen(s) should have been tested simultaneously with the chemical in question. The results should have been reproduced in different laboratories, and a prediction of carcinogenicity should have been confirmed in additional test systems. Confidence in positive results is increased if a mechanism of action can be deduced and if appropriate dose-response data are available. For optimum usefulness, data on purity must be given.

The short-term tests in current use that have been the most extensively validated are the *Salmonella typhimurium* plate-incorporation assay(57-61), the X-linked recessive lethal test in *Drosophila melanogaster*(62), unscheduled DNA synthesis(63) and *in vitro* transformation(61,64). Each is compatible with current concepts of the possible mechanism(s) of carcinogenesis.

An adequate assessment of the genetic activity of a chemical depends on data from a wide range of test systems. The monographs include, therefore, data not only from those already mentioned, but also on the induction of point mutations in other systems(65-70), on structural(71) and numerical chromosome aberrations, including dominant lethal effects(72), on mitotic recombination in fungi(65) and on sister chromatid exchanges(73-74).

The existence of a correlation between quantitative aspects of mutagenic and carcinogenic activity has been suggested(5,72-78), but it is not sufficiently well established to allow general use.

Further information about mutagenicity and other short-term tests is given in references 75-81.

Case reports and epidemiological studies

Observations in humans are summarized in this section.

<u>Summary of Data Reported and Evaluation (Section 4)</u>

Section 4 summarizes the relevant data from animals and humans and gives the critical views of the Working Group on those data.

Experimental data

Data relevant to the evaluation of the carcinogenicity of a chemical in animals are summarized in this section. Results from validated mutagenicity and other short-term tests are reported if the Working Group considered the data to be relevant. Dose-response data are given when available. An assessment of the carcinogenicity of the chemical in animals is made on the basis of all of the available data.

The animal species mentioned are those in which the carcinogenicity of the substance was clearly demonstrated. The route of administration used in experimental animals that is similar to the possible human exposure is given particular mention. Tumour sites are also indicated. If the substance has produced tumours after prenatal exposure or in single-dose experiments, this is indicated.

Human data

Case reports and epidemiological studies that are considered to be pertinent to an assessment of human carcinogenicity are described. Human exposure to the chemical is summarized on the basis of data on production, use and occurrence. Other biological data which are considered to be relevant are also mentioned. An assessment of the carcinogenicity of the chemical in humans is made on the basis of all of the available evidence.

Evaluation

This section comprises the overall evaluation by the Working Group of the carcinogenic risk of the chemical to humans. All of the data in the monograph, and particularly the summarized information on experimental and human data, are considered in order to make this evaluation.

References

1. IARC (1977) IARC Monograph Programme on the Evaluation of the Carcinogenic Risk of Chemicals to Humans. Preamble. IARC intern. tech. Rep. No. 77/002

2. IARC (1978) Chemicals with *sufficient evidence* of carcinogenicity in experimental animals - *IARC Monographs* volumes 1-17. IARC intern. tech. Rep. No. 78/003

3. WHO (1961) Fifth Report of the Joint FAO/WHO Expert Committee on Food Additives. Evaluation of carcinogenic hazard of food additives. WHO tech. Rep. Ser., No. 220, pp. 5, 18, 19

4. WHO (1969) Report of a WHO Scientific Group. Principles for the testing and evaluation of drugs for carcinogenicity. WHO tech. Rep. Ser., No. 426, pp. 19, 21, 22

5. WHO (1974) Report of a WHO Scientific Group. Assessment of the carcinogenicity and mutagenicity of chemicals. WHO tech. Rep. Ser., No. 546

6. WHO (1964) Report of a WHO Expert Committee. Prevention of cancer. WHO tech. Rep. Ser., No. 276, pp. 29, 30

7. IARC (1972) IARC Monographs on the Evaluation of the Carcinogenic Risk of Chemicals to Man, 1, Some Inorganic Substances, Chlorinated Hydrocarbons, Aromatic Amines, *N*-Nitroso Compounds and Natural Products, Lyon, 184 pages

8. IARC (1973) IARC Monographs on the Evaluation of the Carcinogenic Risk of Chemicals to Man, 2, Some Inorganic and Organometallic Compounds, Lyon, 181 pages

9. IARC (1973) IARC Monographs on the Evaluation of the Carcinogenic Risk of Chemicals to Man, 3, Certain Polycyclic Aromatic Hydrocarbons and Heterocyclic Compounds, Lyon, 271 pages

10. IARC (1974) IARC Monographs on the Evaluation of the Carcinogenic Risk of Chemicals to Man, 4, Some Aromatic Amines, Hydrazine and Related Substances, *N*-Nitroso Compounds and Miscellaneous Alkylating Agents, Lyon, 286 pages

11. IARC (1974) IARC Monographs on the Evaluation of the Carcinogenic Risk of Chemicals to Man, 5, Some Organochlorine Pesticides, Lyon, 241 pages

12. IARC (1974) IARC Monographs on the Evaluation of the Carcinogenic Risk of Chemicals to Man, 6, Sex Hormones, Lyon, 243 pages

13. IARC (1974) *IARC Monographs on the Evaluation of the Carcinogenic Risk of Chemicals to Man, 7, Some Anti-thyroid and Related Substances, Nitrofurans and Industrial Chemicals*, Lyon, 326 pages

14. IARC (1975) *IARC Monographs on the Evaluation of the Carcinogenic Risk of Chemicals to Man, 8, Some Aromatic Azo Compounds*, Lyon, 357 pages

15. IARC (1975) *IARC Monographs on the Evaluation of the Carcinogenic Risk of Chemicals to Man, 9, Some Aziridines, N-, S- and O-Mustards and Selenium*, Lyon, 268 pages

16. IARC (1976) *IARC Monographs on the Evaluation of the Carcinogenic Risk of Chemicals to Man, 10, Some Naturally Occurring Substances*, Lyon, 353 pages

17. IARC (1976) *IARC Monographs on the Evaluation of the Carcinogenic Risk of Chemicals to Man, 11, Cadmium, Nickel, Some Epoxides, Miscellaneous Industrial Chemicals and General Considerations on Volatile Anaesthetics*, Lyon, 306 pages

18. IARC (1976) *IARC Monographs on the Evaluation of the Carcinogenic Risk of Chemicals to Man, 12, Some Carbamates, Thiocarbamates and Carbazides*, Lyon, 282 pages

19. IARC (1977) *IARC Monographs on the Evaluation of the Carcinogenic Risk of Chemicals to Man, 13, Some Miscellaneous Pharmaceutical Substances*, Lyon, 255 pages

20. IARC (1977) *IARC Monographs on the Evaluation of the Carcinogenic Risk of Chemicals to Man, 14, Asbestos*, Lyon, 106 pages

21. IARC (1977) *IARC Monographs on the Evaluation of the Carcinogenic Risk of Chemicals to Man, 15, Some Fumigants, the Herbicides 2,4-D and 2,4,5-T, Chlorinated Dibenzodioxins and Miscellaneous Industrial Chemicals*, Lyon, 354 pages

22. IARC (1977) *IARC Monographs on the Evaluation of the Carcinogenic Risk of Chemicals to Man, 16, Some Aromatic Amines and Related Nitro Compounds - Hair Dyes, Colouring Agents and Miscellaneous Industrial Chemicals*, Lyon, 400 pages

23. IARC (1978) *IARC Monographs on the Evaluation of the Carcinogenic Risk of Chemicals to Humans, 17, Some N-Nitroso Compounds*, Lyon, 365 pages

24. IARC (1978) *IARC Monographs on the Evaluation of the Carcinogenic Risk of Chemicals to Humans, 18, Polychlorinated Biphenyls and Polybrominated Biphenyls*, Lyon, 140 pages

25. IARC (1973) *IARC Information Bulletin on the Survey of Chemicals Being Tested for Carcinogenicity*, No. 1, Lyon, 52 pages

26. IARC (1973) *IARC Information Bulletin on the Survey of Chemicals Being Tested for Carcinogenicity*, No. 2, Lyon, 77 pages

27. IARC (1974) *IARC Information Bulletin on the Survey of Chemicals Being Tested for Carcinogenicity*, No. 3, Lyon, 67 pages

28. IARC (1974) *IARC Information Bulletin on the Survey of Chemicals Being Tested for Carcinogenicity*, No. 4, Lyon, 97 pages

29. IARC (1975) *IARC Information Bulletin on the Survey of Chemicals Being Tested for Carcinogenicity*, No. 5, Lyon, 88 pages

30. IARC (1976) *IARC Information Bulletin on the Survey of Chemicals Being Tested for Carcinogenicity*, No. 6, Lyon, 360 pages

31. IARC (1978) *IARC Information Bulletin on the Survey of Chemicals Being Tested for Carcinogenicity*, No. 7, Lyon, 460 pages

32. Hartwell, J.L. (1951) *Survey of Compounds which have been Tested for Carcinogenic Activity*, Washington DC, US Government Printing Office (Public Health Service Publication No. 149)

33. Shubik, P. & Hartwell, J.L. (1957) *Survey of Compounds which have been Tested for Carcinogenic Activity*, Washington DC, US Government Printing Office (Public Health Service Publication No. 149: Supplement 1)

34. Shubik, P. & Hartwell, J.L. (1969) *Survey of Compounds which have been Tested for Carcinogenic Activity*, Washington DC, US Government Printing Office (Public Health Service Publication No. 149: Supplement 2)

35. Carcinogenesis Program National Cancer Institute (1971) *Survey of Compounds which have been Tested for Carcinogenic Activity*, Washington DC, US Government Printing Office (Public Health Service Publication No. 149: 1968-1969)

36. Carcinogenesis Program National Cancer Institute (1973) *Survey of Compounds which have been Tested for Carcinogenic Activity*, Washington DC, US Government Printing Office (Public Health Service Publication No. 149: 1961-1967)

37. Carcinogenesis Program National Cancer Institute (1974) *Survey of Compounds which have been Tested for Carcinogenic Activity*, Washington DC, US Government Printing Office (Public Health Service Publication No. 149: 1970-1971)

38. Pike, M.C. & Roe, F.J.C. (1963) An actuarial method of analysis of an experiment in two-stage carcinogenesis. Br. J. Cancer, 17, 605-610

39. Miller, E.C. & Miller, J.A. (1966) Mechanisms of chemical carcinogenesis: nature of proximate carcinogens and interactions with macromolecules. Pharmacol. Rev., 18, 805-838

40. Miller, J.A. (1970) Carcinogenesis by chemicals: an overview - G.H.A. Clowes Memorial Lecture. Cancer Res., 30, 559-576

41. Miller, J.A. & Miller, E.C. (1976) The metabolic activation of chemical carcinogens to reactive electrophiles. In: Yuhas, J.M., Tennant, R.W. & Reagon, J.D., eds, Biology of Radiation Carcinogenesis, New York, Raven Press

42. Peto, R. (1974) Guidelines on the analysis of tumours rates and death rates in experimental animals. Br. J. Cancer, 29, 101-105

43. Peto, R. (1975) Letter to the editor. Br. J. Cancer, 31, 697-699

44. Hoel, D.G. & Walburg, H.E., Jr (1972) Statistical analysis of survival experiments. J. natl Cancer Inst., 49, 361-372

45. Tomatis, L. (1977) The value of long-term testing for the implementation of primary prevention. In: Hiatt, H.H., Watson, J.D. & Winsten, J.A., eds, Origins of Human Cancer, Book C, Cold Spring Harbor, N.Y., Cold Spring Harbor Laboratory, pp. 1339-1357

46. IARC (1977) Annual Report 1977, Lyon, International Agency for Research on Cancer, p. 94

47. Tomatis, L., Agthe, C., Bartsch, H., Huff, J., Montesano, R., Saracci, R., Walker, E. & Wilbourn, J. (1978) Evaluation of the carcinogenicity of chemicals: a review of the IARC Monograph Programme, 1971-1977. Cancer Res., 38, 877-885

48. Rall, D.P. (1977) Species differences in carcinogenesis testing. In: Hiatt, H.H., Watson, J.D. & Winsten, J.A., eds, Origins of Human Cancer, Book C, Cold Spring Harbor, N.Y., Cold Spring Harbor Laboratory, pp. 1383-1390

49. National Academy of Sciences (NAS) (1975) Contemporary pest control practices and prospects: the report of the Executive Committee, Washington DC

50. Chemical Abstracts Service (1978) Chemical Abstracts Ninth Collective Index (9CI), 1972-1976, Vols 76-85, Columbus, Ohio

51. WHO (1958) Second Report of the Joint FAO/WHO Expert Committee on Food Additives. Procedures for the testing of intentional food additives to establish their safety and use. WHO tech. Rep. Ser., No. 144

52. WHO (1967) Scientific Group. Procedures for investigating intentional and unintentional food additives. WHO tech. Rep. Ser., No. 348

53. Berenblum, I., ed. (1969) Carcinogenicity testing. UICC tech. Rep. Ser., 2

54. Sontag, J.M., Page, N.P. & Saffiotti, U. (1976) Guidelines for carcinogen bioassay in small rodents. Natl Cancer Inst. Carcinog. tech. Rep. Ser., No. 1

55. Committee on Standardized Genetic Nomenclature for Mice (1972) Standardized nomenclature for inbred strains of mice. Fifth listing. Cancer Res., 32, 1609-1646

56. Wilson, J.G. & Fraser, F.C. (1977) Handbook of Teratology, New York, Plenum Press

57. Ames, B.N., Durston, W.E., Yamasaki, E. & Lee, F.D. (1973) Carcinogens are mutagens: a simple test system combining liver homogenates for activation and bacteria for detection. Proc. natl Acad. Sci. (Wash.), 70, 2281-2285

58. McCann, J., Choi, E., Yamasaki, E. & Ames, B.N. (1975) Detection of carcinogens as mutagens in the *Salmonella*/microsome test: assay of 300 chemicals. Proc. natl Acad. Sci. (Wash.), 72, 5135-5139

59. McCann, J. & Ames, B.N. (1976) Detection of carcinogens as mutagens in the *Salmonella*/microsome test: assay of 300 chemicals: discussion. Proc. natl Acad. Sci. (Wash.), 73, 950-954

60. Sugimura, T., Sato, S., Nagao, M., Yahagi, T., Matsushima, T., Seino, Y., Takeuchi, M. & Kawachi, T. (1977) Overlapping of carcinogens and mutagens. In: Magee, P.N., Takayama, S., Sugimura, T. & Matsushima, T., eds, Fundamentals in Cancer Prevention, Baltimore, University Park Press, pp. 191-215

61. Purchase, I.F.M., Longstaff, E., Ashby, J., Styles, J.A., Anderson, D., Lefevre, P.A. & Westwood, F.R. (1976) Evaluation of six short term tests for detecting organic chemical carcinogens and recommendations for their use. Nature (Lond.), 264, 624-627

62. Vogel, E. & Sobels, F.H. (1976) The function of *Drosophila* in genetic toxicology testing. In: Hollaender, A., ed., Chemical Mutagens: Principles and Methods for Their Detection, Vol. 4, New York, Plenum Press, pp. 93-142

63. San, R.H.C. & Stich, H.F. (1975) DNA repair synthesis of cultured human cells as a rapid bioassay for chemical carcinogens. Int. J. Cancer, 16, 284-291

64. Pienta, R.J., Poiley, J.A. & Lebherz, W.B. (1977) Morphological transformation of early passage golden Syrian hamster embryo cells derived from cryopreserved primary cultures as a reliable *in vitro* bioassay for identifying diverse carcinogens. Int. J. Cancer, 19, 642-655

65. Zimmermann, F.K. (1975) Procedures used in the induction of mitotic recombination and mutation in the yeast *Saccharomyces cerevisiae*. Mutat. Res., 31, 71-86

66. Ong, T.-M. & de Serres, F.J. (1972) Mutagenicity of chemical carcinogens in *Neurospora crassa*. Cancer Res., 32, 1890-1893

67. Huberman, E. & Sachs, L. (1976) Mutability of different genetic loci in mammalian cells by metabolically activated carcinogenic polycyclic hydrocarbons. Proc. natl Acad. Sci. (Wash.), 73, 188-192

68. Krahn, D.F. & Heidelburger, C. (1977) Liver homogenate-mediated mutagenesis in Chinese hamster V79 cells by polycyclic aromatic hydrocarbons and aflatoxins. Mutat. Res., 46, 27-44

69. Kuroki, T., Drevon, C. & Montesano, R. (1977) Microsome-mediated mutagenesis in V79 Chinese hamster cells by various nitrosamines. Cancer Res., 37, 1044-1050

70. Searle, A.G. (1975) The specific locus test in the mouse. Mutat. Res., 31, 277-290

71. Evans, H.J. & O'Riordan, M.L. (1975) Human peripheral blood lymphocytes for the analysis of chromosome aberrations in mutagen tests. Mutat. Res., 31, 135-148

72. Epstein, S.S., Arnold, E., Andrea, J., Bass, W. & Bishop, Y. (1972) Detection of chemical mutagens by the dominant lethal assay in the mouse. Toxicol. appl. Pharmacol., 23, 288-325

73. Perry, P. & Evans, H.J. (1975) Cytological detection of mutagen-carcinogen exposure by sister chromatid exchanges. Nature (Lond.), 258, 121-125

74. Stetka, D.G. & Wolff, S. (1976) Sister chromatid exchanges as an assay for genetic damage induced by mutagen-carcinogens. I. *In vivo* test for compounds requiring metabolic activation. Mutat. Res., 41, 333-342

75. Bartsch, H. & Grover, P.L. (1976) Chemical carcinogenesis and mutagenesis. In: Symington, T. & Carter, R.L., eds, Scientific Foundations of Oncology, Vol. IX, Chemical Carcinogenesis, London, Heinemann Medical Books Ltd, pp. 334-342

76. Hollaender, A., ed. (1971a,b, 1973, 1976) Chemical Mutagens: Principles and Methods for Their Detection, Vols 1-4, New York, Plenum Press

77. Montesano, R. & Tomatis, L., eds (1974) Chemical Carcinogenesis Essays, Lyon (IARC Scientific Publications No. 10)

78. Ramel, C., ed. (1973) Evaluation of genetic risk of environmental chemicals: report of a symposium held at Skokloster, Sweden, 1972. Ambio Spec. Rep., No. 3

79. Stoltz, D.R., Poirier, L.A., Irving, C.C., Stich, H.F., Weisburger, J.H. & Grice, H.C. (1974) Evaluation of short-term tests for carcinogenicity. Toxicol. appl. Pharmacol., 29, 157-180

80. Montesano, R., Bartsch, H. & Tomatis, L., eds (1976) Screening Tests in Chemical Carcinogenesis, Lyon (IARC Scientific Publications No. 12)

81. Committee 17 (1976) Environmental mutagenic hazards. Science, 187, 503-514

GENERAL REMARKS ON THE SUBSTANCES CONSIDERED

This eighteenth volume of the *IARC Monographs* is devoted to polychlorinated biphenyls (PCBs) and polybrominated biphenyls (PBBs). Although these two groups of substances are related chemically and biologically, the individual monographs resulted from the deliberations of two separate IARC Working Groups. The monograph on PCBs was prepared by a Working Group which met in Lyon in October, 1977 to consider some *N*-nitroso compounds and PCBs; the monographs on *N*-nitroso compounds were published in May, 1978 as volume 17 of the *Monograph* series. Polybrominated biphenyls were considered by a Working Group which met in Lyon in June, 1978 to consider some halogenated hydrocarbons; those monographs will be published as volume 20 of the *Monograph* series. The names of the scientists who participated in these two Working Groups are listed on pages 5 and 9.

Polychlorinated biphenyls

PCBs are synthetic compounds that have been used for industrial purposes for the past 45 years. Since they do not conduct electricity and are capable of withstanding high temperatures for long periods, they have been used extensively in the electrical industry and in coolant systems. Because of their slow degradation, they have also been used as sealants for wood and cement surfaces. Other uses are as hydraulic fluids, cutting oils and vapour suppressants for insecticide preparations.

Initial concern about the health effects of the various commercial PCB mixtures originated when chloracne and hepatic changes were seen in workers who were involved in the production of these compounds or who were in direct contact with materials containing them (Drinker *et al.*, 1937; Flinn & Jarvik, 1936; Greenburg *et al.*, 1939; Jones & Alden, 1936; Meigs *et al.*, 1954; Schwartz, 1936, 1943; Schwartz & Barlow, 1942). The role played by toxic impurities or breakdown products, such as polychlorinated dibenzofurans and chlorinated naphthalenes, is not known.

Later, Jensen discovered chromatographic peaks equivalent to those of PCBs in tissues from wildlife (Anon., 1966); and following this report, scientists throughout the world detected PCBs in a variety of locations. The full health significance of these compounds came to public attention in 1968, when over 1000 Japanese became ill after being exposed to rice oil contaminated with PCBs. In this incident, a PCB mixture had been used as a coolant, holes had developed in the coolant system, and the PCBs were discharged into the rice oil. The people who consumed it were exposed to an average of about 2 g of the mixture; they subsequently developed chloracne, headache, nausea, vomiting, peripheral numbness, menstrual irregularity, gastrointestinal disturbances and many other nonspecific signs and symptoms (Kuratsune *et al.*, 1972).

Animal studies have shown that PCBs can cross the placental barrier and are excreted in the mother's milk (Barsotti *et al.*, 1976). Irregular menstrual cycles, early abortions and the birth of small, hyperpigmented and hyperkeratotic infants have been observed (Allen & Barsotti, 1976; Barsotti *et al.*, 1976; Kuratsune *et al.*, 1972).

Many humans have detectable levels of PCBs in their tissues and continue to accumulate minute amounts (Kutz & Strassman, 1976; Yobs, 1972). There is a pressing need for additional information on the harmful effects that may result from the long-term, low-level exposure to these compounds that occurs throughout the world.

Recently, a review on the environmental health aspects of PCBs was published (UNEP/WHO, 1976).

Polybrominated biphenyls

PBBs, which are chemically related to PCBs, are solids. They have been used as flame retardants in industrial processes; the main isomer used was the hexabromobiphenyl. As with PCBs, PBBs persist in the environment and can accumulate in body fat.

Attention was first focused on the adverse health effects of PBBs in 1973, following an incident of poisoning in Michigan dairy cattle that had been fed protein-concentrate feeds into which a commercial hexabromobiphenyl had been mixed accidentally in the place of Nutrimaster (magnesium oxide). Widespread human exposure ensued through the consumption of meat, milk and eggs contaminated with PBBs (Kay, 1977).

Since that time, numerous experimental and human investigations have been initiated. Many of the earlier results were reported at a workshop on scientific aspects of polybrominated biphenyls, held at East Lansing, Michigan, in October 1977. The full articles were published as a special volume (volume 23) of Environmental Health Perspectives, in April 1978.

A literature compilation concerning these compounds is available (Winslow & Gerstner, 1978).

References

Allen, J.R. & Barsotti, D.A. (1976) The effects of transplacental and mammary movement of PCBs on infant rhesus monkeys. Toxicology, 6, 331-340

Anon. (1966) Report of a new chemical hazard. New Sci., 32, 612

Barsotti, D.A., Marlar, R.J. & Allen, J.R. (1976) Reproductive dysfunction in rhesus monkeys exposed to low levels of polychlorinated biphenyls (Aroclor 1248). Food Cosmet. Toxicol., 14, 99-103

Drinker, C.K., Warren, M.F. & Bennett, G.A. (1937) The problem of possible systemic effects from certain chlorinated hydrocarbons. J. ind. Hyg. Toxicol., 19, 283-311

Flinn, F.B. & Jarvik, N.E. (1936) Action of certain chlorinated naphthalenes on the liver. Proc. Soc. exp. Biol. (N.Y.), 35, 118-120

Greenburg, L., Mayers, M.R. & Smith, A.R. (1939) The systemic effects resulting from exposure to certain chlorinated hydrocarbons. J. ind. Hyg. Toxicol., 21, 29-38

Jones, J.W. & Alden, H.S. (1936) An acneform dermatergosis. Arch. Dermatol. Syphilol., 33, 1022-1034

Kay, K. (1977) Polybrominated biphenyls (PBB) environmental contamination in Michigan, 1973-1976. Environ. Res., 13, 74-93

Kuratsune, M., Yoshimura, T., Matsuzaka, J. & Yamaguchi, A. (1972) Epidemiologic study on Yusho, a poisoning caused by ingestion of rice oil contaminated with a commercial brand of polychlorinated biphenyls. Environ. Health Perspect., 1, 119-128

Kutz, F.W. & Strassman, S.C. (1976) Residues of polychlorinated biphenyls in the general population of the United States. In: Proceedings of the National Conference on Polychlorinated Biphenyls, Chicago, 1975, EPA-560/6-75-004, Washington DC, US Environmental Protection Agency, pp. 139-143

Meigs, J.W., Albom, J.J. & Kartin, B.L. (1954) Chloracne from an unusual exposure to Aroclor. J. Am. med. Assoc., 154, 1417-1418

Schwartz, L. (1936) Dermatitis from synthetic resins and waxes. Am. J. Public Health, 26, 586-592

Schwartz, L. (1943) An outbreak of halowax acne ('cable rash') among electricians. J. Am. med. Assoc., 122, 158-161

Schwartz, L. & Barlow, F.A. (1942) Chloracne from cutting oils. US Public Health Rep., 57, 1747-1752

UNEP/WHO (United Nations Environment Programme/World Health Organization) (1976) Polychlorinated Biphenyls and Terphenyls. Environ. Health Criteria, 2, Geneva

Winslow, S.G. & Gerstner, H.B. (1978) Polychlorinated Biphenyls, Polybrominated Biphenyls, and their Contaminants: A Literature Compilation, 1967-1977, ORNL/TIRC-78/2, Springfield, Va, National Technical Information Service, US Department of Commerce

Yobs, A.R. (1972) Levels of polychlorinated biphenyls in adipose tissue of the general population of the nation. Environ. Health Perspect., 1, 79-81

POLYCHLORINATED BIPHENYLS

POLYCHLORINATED BIPHENYLS[1]

These substances were considered previously by an IARC Working Group in June 1974 (IARC, 1974). Since that time new data have become available, and these have been incorporated into the monograph and taken into account in the present evaluation.

A number of reviews on polychlorinated biphenyls (PCBs) are available, e.g., Kurnreich *et al.*, 1976, Subcommittee on the Health Effects of Polychlorinated Biphenyls and Polybrominated Biphenyls (1976, 1978), and UNEP/WHO (1976).

1. Chemical and Physical Data

1.1 Synonyms and trade names

The chlorinated biphenyls contain chlorine atoms at the positions indicated by their names.

Biphenyl
Chem. Abstr. Services Reg. No.: 92-52-4
Chem. Abstr. Name: 1,1'-Biphenyl

2-Chlorobiphenyl
Chem. Abstr. Services Reg. No.: 2051-60-7
Chem. Abstr. Name: 2-Chloro-1,1'-biphenyl

4-Chlorobiphenyl
Chem. Abstr. Services Reg. No.: 2051-62-9
Chem. Abstr. Name: 4-Chloro-1,1'-biphenyl

2,2'-Dichlorobiphenyl
Chem. Abstr. Services Reg. No.: 13029-08-8
Chem. Abstr. Name: 2,2'-Dichloro-1,1'-biphenyl

2,3'-Dichlorobiphenyl
Chem. Abstr. Services Reg. No.: 25569-80-6
Chem. Abstr. Name: 2,3'-Dichloro-1,1'-biphenyl

---

[1] Considered by the Working Group in Lyon, October 1977

2,4'-Dichlorobiphenyl

Chem. Abstr. Services Reg. No.: 34883-43-7
Chem. Abstr. Name: 2,4'-Dichloro-1,1'-biphenyl

4,4'-Dichlorobiphenyl

Chem. Abstr. Services Reg. No.: 2050-68-2
Chem. Abstr. Name: 4,4'-Dichloro-1,1'-biphenyl

2,2',3-Trichlorobiphenyl

Chem. Abstr. Services Reg. No.: 38444-78-9
Chem. Abstr. Name: 2,2',3-Trichloro-1,1'-biphenyl

2,2',5-Trichlorobiphenyl

Chem. Abstr. Services Reg. No.: 37680-65-2
Chem. Abstr. Name: 2,2',5-Trichloro-1,1'-biphenyl

2,3',4-Trichlorobiphenyl

Chem. Abstr. Services Reg. No.: none available
Chem. Abstr. Name: 2,3',4-Trichloro-1,1'-biphenyl

2',3,4-Trichlorobiphenyl

Chem. Abstr. Services Reg. No.: 38444-86-9
Chem. Abstr. Name: 2',3,4-Trichloro-1,1'-biphenyl

2,4,4'-Trichlorobiphenyl

Chem. Abstr. Services Reg. No.: 7012-37-5
Chem. Abstr. Name: 2,4,4'-Trichloro-1,1'-biphenyl

2,4',5-Trichlorobiphenyl

Chem. Abstr. Services Reg. No.: 16606-02-3
Chem. Abstr. Name: 2,4',5-Trichloro-1,1'-biphenyl

2,2',3,5'-Tetrachlorobiphenyl

Chem. Abstr. Services Reg. No.: 41464-39-5
Chem. Abstr. Name: 2,2',3,5'-Tetrachloro-1,1'-biphenyl

2,2',4,5'-Tetrachlorobiphenyl

Chem. Abstr. Services Reg. No.: 41464-40-8
Chem. Abstr. Name: 2,2',4,5'-Tetrachloro-1,1'-biphenyl

2,2',5,5'-Tetrachlorobiphenyl

Chem. Abstr. Services Reg. No.: 35693-99-3
Chem. Abstr. Name: 2,2',5,5'-Tetrachloro-1,1'-biphenyl

POLYCHLORINATED BIPHENYLS

2,3,4,4'-Tetrachlorobiphenyl
Chem. Abstr. Services Reg. No.: 33025-41-1
Chem. Abstr. Name: 2,3,4,4'-Tetrachloro-1,1'-biphenyl

2,3',4,4'-Tetrachlorobiphenyl
Chem. Abstr. Services Reg. No.: 32598-10-0
Chem. Abstr. Name: 2,3',4,4'-Tetrachloro-1,1'-biphenyl

2,3',4',5-Tetrachlorobiphenyl
Chem. Abstr. Services Reg. No.: 32598-11-1
Chem. Abstr. Name: 2,3',4',5-Tetrachloro-1,1'-biphenyl

3,3',4,4'-Tetrachlorobiphenyl
Chem. Abstr. Services Reg. No.: 32598-13-3
Chem. Abstr. Name: 3,3',4,4'-Tetrachloro-1,1'-biphenyl

2,2',3,3',6-Pentachlorobiphenyl
Chem. Abstr. Services Reg. No.: 52663-60-2
Chem. Abstr. Name: 2,2',3,3',6-Pentachloro-1,1'-biphenyl

2,2',3,4,5'-Pentachlorobiphenyl
Chem. Abstr. Services Reg. No.: 38380-02-8
Chem. Abstr. Name: 2,2',3,4,5'-Pentachloro-1,1'-biphenyl

2,2',3',4,5-Pentachlorobiphenyl
Chem. Abstr. Services Reg. No.: 41464-51-1
Chem. Abstr. Name: 2,2',3'4,5-Pentachloro-1,1'-biphenyl

2,2',3,4',6-Pentachlorobiphenyl
Chem. Abstr. Services Reg. No.: none available
Chem. Abstr. Name: 2,2',3,4',6-Pentachloro-1,1'-biphenyl

2,2',3,5',6-Pentachlorobiphenyl
Chem. Abstr. Services Reg. No.: 38379-99-6
Chem. Abstr. Name: 2,2',3,5',6-Pentachloro-1,1'-biphenyl

2,2',4,4',5-Pentachlorobiphenyl
Chem. Abstr. Services Reg. No.: 38380-01-7
Chem. Abstr. Name: 2,2',4,4',5-Pentachloro-1,1'-biphenyl

2,2',4,5,5'-Pentachlorobiphenyl
Chem. Abstr. Services Reg. No.: 37680-73-2
Chem. Abstr. Name: 2,2',4,5,5'-Pentachloro-1,1'-biphenyl

2,3,3',4,4'-Pentachlorobiphenyl

Chem. Abstr. Services Reg. No.: 32598-14-4
Chem. Abstr. Name: 2,3,3',4,4'-Pentachloro-1,1'-biphenyl

2,3,3',4',6-Pentachlorobiphenyl

Chem. Abstr. Services Reg. No.: 38380-03-9
Chem. Abstr. Name: 2,3,3',4',6-Pentachloro-1,1'-biphenyl

2,3',4,4',5-Pentachlorobiphenyl

Chem. Abstr. Services Reg. No.: 31508-00-6
Chem. Abstr. Name: 2,3',4,4',5-Pentachloro-1,1'-biphenyl

2,2',3,3',4,6-Hexachlorobiphenyl

Chem. Abstr. Services Reg. No.: 38380-05-1
Chem. Abstr. Name: 2,2',3,3',4,6-Hexachloro-1,1'-biphenyl

2,2',3,3',6,6'-Hexachlorobiphenyl

Chem. Abstr. Services Reg. No.: 38411-22-2
Chem. Abstr. Name: 2,2',3,3',6,6'-Hexachloro-1,1'-biphenyl

2,2',3,4,4',5-Hexachlorobiphenyl

Chem. Abstr. Services Reg. No.: 35694-06-5
Chem. Abstr. Name: 2,2',3,4,4',5-Hexachloro-1,1'-biphenyl

2,2',3,4,4',5'-Hexachlorobiphenyl

Chem. Abstr. Services Reg. No.: 35065-28-2
Chem. Abstr. Name: 2,2',3,4,4',5'-Hexachloro-1,1'-biphenyl

2,2',3',4,5,6'-Hexachlorobiphenyl

Chem. Abstr. Services Reg. No.: 38380-04-0
Chem. Abstr. Name: 2,2',3',4,5,6'-Hexachloro-1,1'-biphenyl

2,2',4,4',5,5'-Hexachlorobiphenyl

Chem. Abstr. Services Reg. No.: 35065-27-1
Chem. Abstr. Name: 2,2',4,4',5,5'-Hexachloro-1,1'-biphenyl

2,2',3,3',4,4',5-Heptachlorobiphenyl

Chem. Abstr. Services Reg. No.: 35065-30-6
Chem. Abstr. Name: 2,2',3,3',4,4',5-Heptachloro-1,1'-biphenyl

2,2',3,3',4,5,6'-Heptachlorobiphenyl

Chem. Abstr. Services Reg. No.: 38441-25-5
Chem. Abstr. Name: 2,2',3,3',4,5,6'-Heptachloro-1,1'-biphenyl

2,2',3,4,4',5,5'-Heptachlorobiphenyl

Chem. Abstr. Services Reg. No.: 35065-29-3
Chem. Abstr. Name: 2,2',3,4,4',5,5'-Heptachloro-1,1'-biphenyl

Synonyms

Chlorinated biphenyl; chlorinated diphenyl; chlorobiphenyl; PCB; PCBs; polychlorinated biphenyl; polychlorobiphenyl

Trade names (The producing country is given in parentheses)

Aroclor (USA); Chlorextol (USA); Clophen (FRG); Dykanol (USA); Fenclor (Italy); Inerteen (USA); Kanechlor (Japan); Noflamol (USA); Phenoclor (France); Pyralene[1] (France); Pyranol (USA); Santotherm (Japan); Sovol (USSR); Therminol[1] (USA)

The approximate molecular compositions of some commercial PCB mixtures are listed in Table 1.

1.2 Structural and molecular formulae and molecular weight

The structural formula of the unsubstituted biphenyl with the numbering of the carbon atoms in the ring is given below.

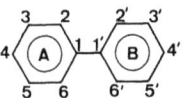

Chlorobiphenyl	$C_{12}H_9Cl$	Mol. wt:	188.7
Dichlorobiphenyl	$C_{12}H_8Cl_2$	"	223.1
Trichlorobiphenyl	$C_{12}H_7Cl_3$	"	257.6
Tetrachlorobiphenyl	$C_{12}H_6Cl_4$	"	292.0
Pentachlorobiphenyl	$C_{12}H_5Cl_5$	"	326.4
Hexachlorobiphenyl	$C_{12}H_4Cl_6$	"	360.9
Heptachlorobiphenyl	$C_{12}H_3Cl_7$	"	395.3
Octachlorobiphenyl	$C_{12}H_2Cl_8$	"	329.7
Nonachlorobiphenyl	$C_{12}HCl_9$	"	364.1
Decachlorobiphenyl	$C_{12}Cl_{10}$	"	398.5

In all, there are 209 possible PCB isomers (Fig. 1).

---

[1] These products, as now formulated, no longer contain PCBs.

Table 1

Approximate weight percentages of chlorobiphenyls in some commercial PCB mixtures[a]

Number of chlorines	Aroclors[b]							Kanechlors			Fenclors Fenclor (Denmark)
	1221[c]	1232	1016	1242[c]	1248	1254[c]	1260[d]	KC-300	KC-400	KC-500	
0	11	6	Tr	Tr	–	Tr	–	–	–	–	–
1	51	26	1	1	–	Tr	–	–	–	–	–
2	32	29	20	17	1	0.5	–	17	3	–	–
3	4	24	57	40	23	1	–	60	33	–	–
4	2	15	21	32	50	21	–	23	44	5	–
5	0.5	0.5	1	10	20	48	12	0.6	16	27	–
6	–	–	Tr	0.5	1	23	46	–	5	55	–
7	–	–	–	Tr	–	6	35	–	–	13	–
8	–	–	–	–	–	1	6	–	–	–	–
9	–	–	–	–	–	–	–	–	–	–	–
10	–	–	–	–	–	–	–	–	–	–	100

[a] From Nisbet (1976a)
[b] The Aroclor number designations consist of two parts: the first two digits, 12, indicate that the molecule is a biphenyl (earlier products, no longer available, were represented by 54 for terphenyls and 25 and 44 for blends of biphenyls and terphenyls). The second two digits, 32, 42, 54, etc., represent the approximate weight per cent of chlorine in the product. The only exception to this is Aroclor 1016, which is a biphenyl containing 41% chlorine by weight.
[c] Information derived from several sources
[d] Phenoclor DP6 (from France) and Clophen A60 (from the Federal Republic of Germany) are almost identical in composition to Aroclor 1260.
Tr = trace

Figure 1[a]

Numbers of combinations, according to possible distributions of chlorine atoms in the two rings of the biphenyl

		No. of chlorine atoms on ring A					
		0	1	2	3	4	5
No. of chlorine atoms on ring B	0	1	3	6	6	3	1
	1		6	18	18	9	3
	2			21	36	18	6
	3				21	18	6
	4					6	3
	5						1

[a]From Hutzinger et al., 1974a

1.3 Chemical and physical properties

The outstanding physical and chemical characteristics of PCBs are their thermal stability, resistance to oxidation, acids, bases and other chemical agents, as well as their excellent dielectric (insulating) properties. Chlorobiphenyls are colourless crystals when isolated in pure form by recrystallization from suitable solvents. Commercial products are liquids because the melting point is depressed when PCB components are mixed. The solubility of PCBs in water, an important physical property from an environmental point of view, is extremely low; it ranges from 0.007-5.9 mg/l for the chlorobiphenyl isomers that occur commonly (Hutzinger et al., 1974a). However, all PCBs are soluble in oils and organic solvents. Other physical properties of 38 PCBs are given in Table 2; the physical properties of some Aroclors are given in Table 3. Only those chlorobiphenyls that have been properly characterized and are present in significant concentrations in commercial PCB formulations are listed.

IARC MONOGRAPHS VOLUME 18

Table 2

Physical properties of PCBs

PCB	MP (°C)^a	λ_{max} nm $(E_1^1)^{a,b}$
Monochlorobiphenyls		
2-	34	204; 240 (2074; 539)
4-	77.7	199; 253 (2290; 1085)
Dichlorobiphenyls		
2,2'-	60.5	208; 230; 273 (1614; 296; 24)
2,3'-		
2,4'-	44.5-46	
4,4'-	149-150	200; 258 (1878; 1026)
Trichlorobiphenyls		
2,2',3-	28.1-28.8	
2,3',4-		210; 246 (1743; 466)
2,4,4'-	57-58	205; 250 (1650; 575)
2,2',5-	43-44	197; 267; 275; 283 (2426; 47.7; 45.4; 31.8)
2,4',5-	67	
2',3,4-	65-66	
Tetrachlorobiphenyls		
2,3,4,4'-	142	250 (432)
2,2',3,5'-	47	266.5^c; 275.5; 283 (32; 40; 34)
2,2',4,5'-	66-68.5	274; 281.5 (41; 34)
2,3',4,4'-	127-127.5	253 (545)
2,2',5,5'-	86-89	204; 214^c; 276; 284 (1483; 1195; 45.2 42.8)
2,3',4',5-	104	247; 285 (555; 86)
3,3',4,4'-	182-184	261 (507)
Pentachlorobiphenyls		
2,3,3',4,4'-	101-105	253 (77)
2,2',3,4,5'-	111.5-113	
2,2',3,3',6-		
2,2',3,4',6-		
2,2',3,5',6-		270; 277; 284 (25.5; 37.7; 36)
2,3,3',4',6-		274^c; 281.5^c (46.4; 42.3)
2,2',3',4,5-	81-82	272; 281; 289.5 (13.4; 15.4; 12.8)
2,2',4,4',5-		
2,2',4,5,5'-	76.5-77.5	275.5^c; 281; 289 (40; 53; 40)
2,3',4,4',5-	105-107	253 (77)
Hexachlorobiphenyls		
2,2',3,4,4',5-	77-78	
2,2',3,3',4,6-		
2,2',3,4,4',5'-	78.5-80	
2,2',3,3',6,6'-	114-114.5	
2,2',4,4',5,5'-	103-104	211; 282; 290 (1260; 44.3; 31)
2,2',3',4,5,6'-		
Heptachlorobiphenyls		
2,2',3,3',4,4',5-	134.5-135.5	
2,2',3,3',4,5,6'-	130.5-130.7	
2,2',3,4,4',5,5'-	109-110	

^{a}From Hutzinger et al., 1974a
^{b}From Sundström, 1973
^{c}Inflexion

Infra-red, nuclear magnetic resonance (^{1}H-NMR or ^{13}C-NMR) and mass spectra have been given by Hutzinger et al., 1974a; Safe & Hutzinger, 1972; Tas & Kleipool, 1972; Tas & de Vos, 1971; Webb & McCall, 1972; Welti & Sissons, 1972 (^{1}H-NMR); and Wilson, 1975 (^{13}C-NMR).

Table 3

Physical properties and characteristics of some Aroclors[a]

Property	1221	1016	1242	1254
Appearance	Clear, mobile oil	Clear, mobile oil	Clear, mobile oil	Light-yellow, viscous liquid
Specific gravity	1.182–1.192 (25/15.5°C)	1.362–1.372 (25/15.5°C)	1.381–1.392 (25/15.5°C)	1.495–1.505 (65/15.5°C)
Viscosity at 37.8°C (Saybolt Universal Seconds)	38–41	71–81	82–92	1800–2500
Refractive index (n^{20})	1.617–1.618 (20°C)	1.6215–1.6235 (25°C)	1.627–1.629 (20°C)	1.6375–1.6415 (25°C)
Pour-point (°C)	1 (crystals)	–14 (max)	–19	10
Acidity (mg KOH/g, max)	0.014	0.010	0.015	0.010
Moisture (ppm, max)	–	35	50	50
Distillation range (°C, corrected)	275–320	323–356	325–366	365–390
Vaporization rate at 100°C (g/cm²/h)	0.00174	–	0.000338	0.000053

[a] From Hutzinger et al., 1974a and Monsanto Industrial Chemicals Co., 1975

The PCBs are chemically very inert and are stable to conditions of hydrolysis and oxidation encountered in industrial use. Photochemical degradation may be one route of their breakdown in the environment: photochemical experiments conducted under simulated natural conditions on a number of pure chlorobiphenyls and on commercial PCB preparations have indicated several degradative reactions, such as dechlorination, polymerization and solvolysis (Hutzinger et al., 1972a).

Heating of Aroclor 1248 at 300°C in an atmosphere of oxygen has been found to yield about 1000 ppm polychlorodibenzofurans (PCDFs). Major components were dichloro- and trichlorodibenzofurans (Morita et al., 1978).

For data on the persistence of these compounds in the environment, see section 2.2.

1.4 Technical products and impurities

Contamination of PCB mixtures with polychlorinated dibenzofurans (PCDFs) and polychlorinated naphthalenes:

Two reviews on PCDFs have been published recently (Buser, 1978; IARC, 1978a).

Vos et al. (1970) were able to identify tetrachlorodibenzofurans, pentachlorodibenzofurans as well as chlorinated naphthalenes in samples of Phenoclor DP-6 and Clophen A-60 (both 60% chlorine), but not in a sample of Aroclor 1260. Bowes et al. (1975) examined samples of Aroclor 1248, 1254 and 1260 produced in 1969, samples of Aroclor 1254 from 1970 and Aroclor 1016 from 1972 and the samples of Aroclor 1260, Phenoclor DP-6 and Clophen A-60 that had previously been analysed by Vos et al. (1970); they found PCDFs in all Aroclor preparations except Aroclor 1016 and in Clophen A-60 and Phenoclor DP-6 (Table 4).

Nagayama et al. (1976) found detectable levels (1-17 µg/g) of PCDFs in samples of Kanechlor (see Table 5); Roach & Pomerantz (1974) found varying concentrations of PCDFs in Kanechlor 400.

Nagayama et al. (1976) found about 5 µg/g PCDFs in 3 samples of 'Yusho oil' that were contaminated with Kanechlor 400; the major constituents were tetra- and pentachlorodibenzofurans. Rappe et al. (1977) recently confirmed by capillary gas chromatography-mass fragmentography that 2,3,7,8-tetrachlorodibenzofuran was one of the main PCDFs in 'Yusho oil'.

Table 4

Concentrations of chlorinated dibenzofurans[a] in Aroclor, Clophen and Phenoclor[b]

PCB	4-Cl	5-Cl	6-Cl	Total
Aroclor 1248 (1969)	0.5(25)	1.2(60)	0.3(15)	2.0
Aroclor 1254 (1969)	0.1(6)	0.2(12)	1.4(82)	1.7
Aroclor 1254 (1970)	0.2(13)	0.4(27)	0.9(60)	1.5
Aroclor 1260 (1969)	0.1(10)	0.4(40)	0.5(50)	1.0
Aroclor 1260 (lot AK3)	0.2(25)	0.3(38)	0.3(38)	0.8
Aroclor 1016 (1972)	ND	ND	ND	-
Clophen A-60	1.4(17)	5.0(59)	2.2(26)	8.4
Phenoclor DP-6	0.7(5)	10.0(74)	2.9(21)	13.6

ND = not detected (<0.001 µg/g)
[a] Expressed as µg/g PCB. Values in parentheses represent quantity as percentage of total dibenzofurans.
[b] From Bowes et al. (1975)

Table 5

Concentrations of chlorinated dibenzofurans in Kanechlors[a]

Kanechlor	Chlorodibenzofurans						Concentration (µg/g)	
	Di-	Tri-	Tetra-	Penta-	Hexa-	Hepta-	b	c
300			+	+			1	1.5
400	+	+	+	+			18	17
500		+		+	+	+	4	2.5
600			+	+	+	+	5	3

[a] From Nagayama et al. (1976)
[b] Calculated from peak heights
[c] Calculated by perchlorination method

2. Production, Use, Occurrence and Analysis

Three reviews on PCBs have been published (Hutzinger et al., 1974a; Peakall & Risebrough, 1975; US Environmental Protection Agency, 1976).

2.1 Production and use

(a) Production

Dichlorodiphenyl was first prepared by Griefs (1867) by heating 4,4'-biphenylbis(diazonium) platinum chloride salt with sodium carbonate. PCBs are prepared industrially by the chlorination of biphenyl with anhydrous chlorine in the presence of a catalyst such as iron filings or ferric chloride. The products are complex mixtures of chlorobiphenyls, whose degree of chlorination depends principally on the time of contact (12-36 h) of the biphenyl with anhydrous chlorine. The crude product is blown with air, and a small amount of lime is added to remove hydrogen chloride and ferric chloride. The resulting chlorinated mixtures are batch-distilled to remove colour and traces of hydrogen chloride and ferric chloride (US Environmental Protection Agency, 1976).

PCBs have been produced commercially in the US since 1929 (Interdepartmental Task Force on PCBs, 1972). In 1964, six US companies had registered trademarks for commercial brands of PCBs (Hubbard, 1964); however, there is currently only one US producer (Durfee, 1976; US Environmental Protection Agency, 1976).

It has been estimated that some 450 million kg PCBs had been sold in North America by 1970 (Interdepartmental Task Force on PCBs, 1972). Annual US production of PCBs increased from 17.2 million kg in 1960 to 38.5 million kg in 1970, when production reached its maximum. Total US sales in 1970 amounted to 33.2 million kg and comprised 22.1 million kg Aroclor 1242, 5.6 million kg Aroclor 1254, 2.2 million kg Aroclor 1260, 1.8 million kg Aroclor 1248, 670 thousand kg Aroclor 1221, 464 thousand kg Aroclor 1262, 150 thousand kg Aroclor 1268 and 118 thousand kg Aroclor 1232. Aroclor 1016 was first sold in 1971. By 1974, total US production had dropped to 18.4 million kg; it was 3.9 million kg for the first quarter of 1975. US sales in 1974 were of 15.6 million kg, including 10 million kg Aroclor 1016, 2.8 million kg each of Aroclor 1242 and Aroclor 1254 and 25.9 thousand kg Aroclor 1221 (US Environmental Protection Agency, 1976). The sole US producer planned to withdraw completely from the production of PCBs by the end of October 1977 (Anon., 1977).

US imports in 1976 were 132 thousand kg (US International Trade Commission, 1977). Exports in 1974 amounted to 2.4 million kg (US Environmental Protection Agency, 1976); about 80-90% of these were decachlorobiphenyls from Italy, and the other 10-20% came from France (Durfee, 1976).

PCBs were first produced commercially in Japan in 1954; by the end of 1971, a total of 57.3 million kg had been produced. Japanese production in

1970 amounted to more than 11 million kg. In 1971, 170 thousand kg were imported, and 730 thousand kg were exported. PCB production in Japan was suspended in 1972 (Organisation for Economic Cooperation and Development, 1976).

In 1973, France reported production of 9.7 million kg PCBs, Italy reported production of 2.5 million kg, and the UK reported production of 4.1 million kg. Austria, Czechoslovakia, the Federal Republic of Germany, Spain, and the USSR also produced PCBs in 1973, but the quantities are not known (Hutzinger et al., 1974a; US Environmental Protection Agency, 1976).

The following countries reported imports of PCBs in 1973: Canada, 1.1 million kg; Finland, 241 thousand kg; France, 300 thousand kg; Italy, 1.6 million kg; New Zealand, 18 thousand kg; Norway, 18 thousand kg; Sweden, 318 thousand kg; and the UK, 4.5 thousand kg. PCBs were exported in 1973 by: France, 4.6 million kg; Italy, 1.1 million kg; and the UK, 3.8 million kg (US Environmental Protection Agency, 1976).

(b) Use

The physical properties of PCBs (see Table 3) prompted their use in numerous industrial products prior to 1972. Because of their good thermal and chemical stability and dielectric properties they were used in 'nominally closed' applications, such as in heat transfer and hydraulic fluids and lubricants, and in 'open end' applications (where emission of PCBs into the environment is more probable since it is not controlled), such as in plasticizers, surface coatings, inks, adhesives, pesticide extenders and for microencapsulation of dyes for carbonless duplicating paper (US Environmental Protection Agency, 1976).

PCBs have been used in immersion oil for microscopes (Alvares et al., 1974).

They were also used as a catalyst carrier in the polymerization of olefins, in the conversion of water-permeable solids to non-permeable states and combined with insecticide and bactericide formulations. Mixtures of PCBs and chlorinated naphthalenes were also used to insulate electric wires and cables in the mining and shipbuilding industries.

Between 1957 and 1971, the US producer manufactured 12 different types of Aroclors, with chlorine contents ranging from 21-68%. Aroclor 1242 and grades with less than 42% chlorine made up about 48% of total consumption during this period. The current and former US uses of Aroclors are summarized by type in Table 6. The 1973 use patterns in countries that are members of the Organisation for Economic Cooperation and Development (OECD) are summarized in Table 7 (US Environmental Protection Agency, 1976).

An estimated 180 thousand kg decachlorobiphenyl that were imported into the US in 1974 from Italy were used as filler for investment casting waxes (Durfee, 1976).

Table 6

End uses of Aroclors by type[a]

End use	1016	1221	1232	1242	1248	1254	1260	1262	1268
Current									
Capacitors	x								
Transformers		x		x[b]		x[b]	x[b]		
				x		x			
Former									
Heat transfer									
Hydraulics/lubricants			x	x	x	x	x		
Hydraulic fluids					x	x			
Vacuum pumps		x		x					
Gas-transmission turbines									
Plasticizers									
Rubbers		x	x	x	x	x			x
Synthetic resins					x	x			x
Carbonless paper				x					
Miscellaneous									
Adhesives		x	x	x	x	x	x	x	x
Wax extenders				x		x			
Dedusting agents						x			
Inks						x	x		
Cutting oils						x			
Pesticide extenders						x			
Sealants and caulking compounds									

[a] From US Environmental Protection Agency (1976)
[b] Discontinued

Table 7

Use of PCBs (in million kg) by OECD member countries (1973)[a]

Country	Trans-formers	Capacitors (large)	Capacitors (small)	Heat-transfer systems	Hydraulic equipment	Vacuum pumps	Lubricating and cutting oils	Plasti-cizers	Others
Australia									
Austria									
Belgium	.53	.18	—	?	?	?	?	?	0
Canada	.90[b]	.20[c]	.19[c]	0	0	0	0	?	0
Denmark									
Eire									
Federal Republic of Germany									
Finland	.04[b]	.20[c]	0	0	0	0	?	0	0
France	2.94	1.30		.07	.08	.005	.22	.66	.11
Greece									
Iceland									
Italy	1.23	1.43[b,c]	0	0	0	0	0	.29[c,d]	.07[e,f]
Japan									
Luxembourg	.03	0	0	0	0	.002	0	.009	0
New Zealand	0	.02[c]	.002	0	0	0	0	0	0
Norway									
Portugal									
Spain	0	.33	0	0	0	0	0	0	0
Sweden									
Switzerland	?	?	?	?	?	?	?	0	0
The Netherlands									
Turkey						.02		?	.02[i]	?
UK	.32[b]	0.83[g]							
US	17.19[h]								

[a] From US Environmental Protection Agency (1976)
[b] Containing 54 wt % Cl
[c] " " 42 "
[d] " " 64 "
[e] " " 70 "
[f] Used as a fire-retardant in plastics
[g] This figure includes about 6% from previously imported material.
[h] This figure includes 0.059 million kg of imported material.
[i] Used in investment casting

All US use of PCBs in 1974 was in closed systems, for the fabrication of capacitors and transformers; about 70% is now used in capacitors and 30% in transformers. In the US, about 55% of transformers contain PCBs (most others contain mineral oil), and these are used only when safety and reliability are of prime importance. Life expectancy for transformers containing PCBs is more than 30 years, and over 135 000 have been put into service since 1932. About 95% of the capacitors made in the US contain PCBs, and about 100 million of these are made yearly. The life expectancy for capacitors used in lighting is more than 10 years, and for those used in electrical utilities, over 20 years (Durfee, 1976).

The total PCB content of all heavy-duty capacitors in Sweden has been estimated to be 2.5 million kg.

In Japan, the total volume of PCBs used during the period 1954-1972 was 54 million kg. An estimated 37 million kg were used by the electrical industry for such items as transformers and condensers, 8.6 million kg were used as heat-transfer agents, 5.4 million kg for non-carbon copy paper and 2.9 million kg for miscellaneous purposes such as paints and lubricants (Tatsukawa, 1976).

As a result of the health implications of exposure to PCBs, Japan prohibited its production in 1972. Manufacturers in the UK may sell PCBs only for their use as dielectric fluids; manufacturers in the Federal Republic of Germany may sell them only as dielectric, heat-transfer and hydraulic fluids. Since 1972, Switzerland and Sweden have restricted sales of PCBs to heavy industrial use, which is subject to prior authorization. The sole manufacturer of PCBs in the US indicated that the sale of PCBs in 1971 was limited to use in closed systems; sales for heat-transfer application were discontinued in 1972, and they planned to discontinue production of PCBs at the end of 1977.

On 2 February 1977, final regulations were issued in the US prohibiting PCB manufacturers and producers of transformers and capacitors from discharging PCBs into waterways. This regulation also limits the PCB level in ambient water to 0.001 µg/l (US Environmental Protection Agency, 1977a).

On 24 May 1977, the US Environmental Protection Agency proposed regulations that prescribed requirements for the disposal and marking of PCBs and articles and equipment that contain them. The regulation calls for high-temperature concentration of all liquid PCBs, which are defined as any liquid containing 0.05% (500 ppm) or more PCBs (US Environmental Protection Agency, 1977b).

Permissible levels of PCB compounds in the working environment have been established in various countries. The US Occupational Safety and Health Administration health standards require that an employee's exposure to PCBs containing 42% chlorine shall at no time exceed an 8-hour time-weighted average of 1 mg/m^3, and exposure to PCBs containing 54% chlorine shall not exceed an 8-hour time-weighted average of 0.5 mg/m^3 over any eight-hour shift of a 40-hour work week (US Occupational Safety and Health Administration, 1977). The maximum allowable concentrations for these two products in the USSR is 1.0 mg/m^3; in the Federal Republic of Germany, 1.0 and 0.5 mg/m^3, respectively; in the German Democratic Republic, 1.0 mg/m^3; in Sweden, 0.5 mg/m^3; and in Czechoslovakia, 1.0 and 0.5 mg/m^3, respectively (Winell, 1975).

On 1 April 1977, the US Food and Drug Administration established tolerance levels in various foods in an attempt to reduce human exposure to PCBs: in milk and dairy products (fat basis), they were reduced from 2.5 to 1.5 µg/g; in poultry (fat basis), from 5 to 3 µg/g; in eggs, from 0.5 to 0.3 µg/g; and in fish and shellfish, from 5 to 2 µg/g. The following tolerance levels were left unchanged: in finished animal feed, 0.2 µg/g; in animal feed components, 2.0 µg/g; in infant and junior foods, 0.2 µg/g; and in paper food-packaging materials without a PCB-impermeable barrier, 10 µg/g (US Food and Drug Administration, 1977a).

In Japan, the tolerance level is 3 µg/g in in-shore fish and 0.5 µg/g in off-shore fish (Fujiwara, 1975). In Sweden, fish containing levels in excess of 5 µg/g are banned from sale.

2.2 Occurrence

(a) Occupational exposure

The level of PCB exposure experienced by workers varies considerably. Elkins (1959) reported that maximum PCB levels in the workroom air in several plants in the US ranged from 0.2-10.5 mg/m^3. In a study of an outbreak of dermatitis caused by PCBs, the concentration in the area 4 months prior to the outbreak was determined to have been 0.1 mg/m^3 (Meigs et al., 1954). Analyses performed between 1953 and 1957 in a Japanese factory making capacitors showed that the PCB concentration in the air was 0.4-6.7 mg/m^3 (Subcommittee on the Health Effects of Polychlorinated Biphenyls and Polybrominated Biphenyls, 1976).

Karppanen & Kolho (1973) reported that 11 'healthy' workers employed for 4 years in a capacitor factory in Finland, where Aroclor 1242 had been used as the impregnating fluid, had a concentration of PCBs in their blood (0.07-1.9 µg/g) that was approximately 50 times greater than that of a control group with no particular exposure to PCBs (0.003-0.012 µg/g). The PCB pattern in the exposed workers differed markedly from the pattern of the PCB used; small amounts of the components contained in Aroclor 1254 were present in the Aroclor 1242 used. Consequently, the total PCB intake must have been higher than that reflected by the levels detected in the body; the average PCB concentrations in the air did not exceed 'internationally accepted limits'.

The PCB concentration in the blood of the workers 2 years after the use of PCBs in the production of carbonless copy paper had ceased was 'approximately' 0.01-0.02 µg/g, which was 10% of the levels found during the period when PCBs were used. The PCBs found in the workers' blood contained one more chlorine atom than did those handled in the factory (Hasegawa et al., 1973). A stockroom for carbonless copy paper was found to contain 0.07 mg/m^3 of air (Tatsukawa, 1976).

Ouw et al. (1976) reported that workers in the electrical industry in New South Wales were exposed to air carrying from 0.32-2.22 mg/m^3 Aroclor 1242, with a mean of 1.27 mg/m^3, and they had approximately 0.4 µg/kg PCBs in their blood. Contact with the PCBs was primarily via the skin.

Nine samples of a widely used microscopic immersion oil marketed in Europe, North America and Japan were found to contain 30-45% PCBs (Bennett & Albro, 1973). Tests run on fluorescent light ballasts by the US Environmental Protection Agency indicate that PCBs are emitted during burnout of certain types of ballasts. Samples of air were taken at various distances from a spent ballast, starting five minutes after the burnout: the highest level of PCB emission, 0.166 mg/m^3, occurred after 4.5 hours, at a point one metre below the burned-out ballast; the lowest concentration, 0.012 mg/m^3, occurred after 4.5 hours at a distance of 4.5 metres from the fixture (Staiff et al., 1974).

It has been estimated that 12 000 persons in the US are occupationally exposed to PCBs (Lloyd et al., 1976; NIOSH, 1977).

(b) Air

In a Japanese survey, it was determined that the urban areas of large cities, such as Tokyo, contained 0.02 µg/m^3 PCBs, while medium-sized cities, such as Matsuyama, had 0.002-0.005 µg/m^3. The air around electrical appliance factories and paper recycling mills had much higher PCB levels (up to 12 µg/m^3). The PCB isomers found in the air often had a lower chlorine content than those in the volatilization source (Tatsukawa, 1976; Tatsukawa & Watanabe, 1972).

Kutz & Yang (1976) found that the average concentration of PCBs in samples of ambient air from suburban locations in Miami (Florida), Fort Collins (Colorado) and Jackson (Mississippi) in the US, in April, May and June 1975 was 0.1 µg/m^3.

Young et al. (1976) estimated that the fallout of PCBs from the air in southern California was 1400 kg/yr in the spring, 1430 kg/yr in the summer and 2700 kg/yr in the autumn. Nisbet (1976b) took into consideration that snowmelt contains 200 ng/kg and the air 100 ng/m^3 and that dry deposition accounts for 70-200 (and locally up to 2000) ng/m^3/day PCBs. He estimated that 1 million kg PCBs fall on the US yearly in rain and particulate matter.

(c) Water and sediments

A survey in southern California showed that 5400 kg PCBs were discharged annually into the municipal wastewater, approximately 800 kg/yr occurred as a result of runoff, and less than 250 kg/yr resulted from industrial discharge (Young et al., 1976).

A national survey conducted in the US showed that the PCB levels in unfiltered water samples during 1971-1974 ranged between 0.1 and 3.0 µg/l (Dennis, 1976). Detectable levels of PCBs were recorded in the Milwaukee River in Wisconsin, which flows into Lake Michigan; the major sources of the PCBs were municipal and industrial effluents (Veith & Lee, 1971). Kleinert (1976) estimated that the waters of Lake Michigan contain as much as 10 ng/l PCBs.

In Japan, the discharge of PCBs into lakes, streams and the ocean occurs primarily as a result of municipal and industrial discharge. In a survey of the surface water of Suruga Bay, near Fuji, where there are a large number of paper mills which use recycled paper, the levels of PCBs varied from 910-1600 ng/l. Surface water in Tokyo Bay in 1973 contained as much as 320 ng/l. Ocean water off Tokyo Bay was found to contain PCBs at a concentration of approximately 3.0 ng/l at a depth of 20 metres; at 200 metres the level was nearly 3.5 ng/l, and at 1500 metres it was approximately 0.5 ng/l (Fukushima, 1974).

Harvey et al. (1974) reported that the PCB concentrations in North Atlantic surface water declined by 40-fold during the period 1972-74, from 41 ng/l to less than 1 ng/l; but this result has been questioned by Longhurst & Radford (1975). In another investigation, Eichner (1976) found that the PCB levels in the Rhine River and Lake Constance had increased slightly during the period 1973-75, from 10 to 75 ng/l.

The average concentration of PCBs in samples of seawater taken from 11 stations off the northwestern Mediterranean coast in 1975 was 13 ng/l (Elder, 1976).

Nadeau & Davis (1976) found up to 2.8 mg/l PCBs in the water and up to 6700 mg/kg in the sediments of the Hudson River in the vicinity of a factory using PCBs. Dennis (1976) reported that bottom sediments from major drainage basins throughout the US and Puerto Rico contained from 1.2-160 µg/kg PCBs during the years 1971-1974. The highest level was found in the basin east of the Mississippi River. Sediment samples in the Milwaukee River in Wisconsin contained 3.5 mg/kg PCB (Kleinert, 1976).

Sediments in bodies of water from 1445 sites throughout Japan contained less than 1 mg/kg PCBs. However, sediments in the rivers entering Tokyo Bay, where many factories using PCBs are located, contained up to 2.7 mg/kg (Tatsukawa, 1976; Tatsukawa & Watanabe, 1972). The content of PCBs tended to be higher in finer-grained sediments: those <74 µ in size contained 117 mg/kg, those >74 µ contained 1-4 mg/kg (Murakami & Takeishi, 1977).

(d) Soil

In a survey of agricultural soils throughout the US in 1972, only 0.1% of samples contained detectable levels of PCBs. In urban areas, the frequency and levels of PCBs were higher: 12 of 19 soil samples from metropolitan areas (63%) showed detectable levels. A PCB pattern corresponding to 54% chlorine was identified in 40% of the positive samples, and a pattern corresponding to 60% was recorded in 20% of the samples (Carey & Gowen, 1976).

In Japan, a nationwide survey of soil samples from 88 sites showed that 40% of samples had less than 0.01 µg/g PCB, 24% had from 0.01-0.10 µg/g, 21% had from 0.11-1.0 µg/g, 7% had from 1.1-10 µg/g, 3% had from 10.1-100 µg/g and 5% had more than 100.1 µg/g. As a result of contamination of soil and the atmosphere with PCBs, detectable levels have been recorded in vegetation: an analysis of PCBs in unhulled rice from 33 sites in Japan in 1972 showed that 79% had less than 0.01 µg/g, 15% had from 0.01-0.10 µg/g, 3% had from 0.11-1.0 µg/g and 3% had from 1.1-10.0 µg/g (Tatsukawa, 1976).

Samples containing up to 300 µg/g PCBs have been taken from soils which were fertilized with sewage sludge containing PCBs or subject to overflow from streams heavily contaminated with PCBs. Soils fertilized with PCB-containing sewage sludge contained 8.5 µg/g, and grass grown on it contained 1.16 µg/g PCBs; beets picked up 0.6 µg/g when grown in soil containing 4 µg/g PCBs (Jordan, 1977).

(e) Food

The majority of people are exposed to PCBs *via* the diet. One survey of food purchased in a local market in Japan indicated that a total PCB intake from prepared food varied from 3.8-50 µg/person/day. An average of 80% of the weekly intake of PCBs was from fish, 6% from meat and 6% from eggs (Tatsukawa, 1976).

Jelinek & Corneliussen (1976) showed that fish, cheese, eggs and by-products used in animal feed were the main commodities in the US that were contaminated with PCBs. The PCB content of all food items except fish decreased between 1971-1975: it was estimated that the PCB content of a teenage boy's diet was 15 µg/day in 1971 and 8.7 µg/day in 1975.

Fish accumulate PCBs to more than 100 000 times the level present in water (Tatsukawa, 1976). In Japan, in 1972, more than 1 µg/g was found in the fresh, edible parts of 16% of sea-water fish and in 18% of fresh-water fish. Fish from the most highly polluted localities contained more than 3 µg/g PCBs (Fujiwara, 1975).

In a 1972-1974 survey in Lake Michigan, trout contained from 12.9-22.9 µg/g, coho salmon contained from 10.4-12.2 µg/g, and bloaters had from 5.2-5.7 µg/g (Willford *et al.*, 1976).

Northern common shiner and northern rock bass collected in the Hudson River near a PCB-handling factory contained 78 µg/g and 350 µg/g, respectively (Nadeau & Davis, 1976). Levels of PCBs in Canadian commercial marine fish in 1975 (samples based on the portion of the fish that would normally be consumed) ranged from 0.07-2.65 µg/g, and levels in fresh-water fish ranged from 0.10-17.14 µg/g (Graham, 1976). Fish in the Danube River and Lake Hallstatt in Austria collected between 1973 and 1975 had PCB residue levels of 0.1-0.3 µg/g (Zislavsky, 1976).

Livers from fish caught in a Swedish fjord contained approximately 5 µg/g PCBs, while those caught in the open sea contained 0.2 µg/g (Falkmer et al., 1978).

(f) Animals

Levels of PCBs found in starlings by the National Fish and Wildlife Monitoring Program ranged from 0.006-1.88 µg/g (wet weight), with a mean value of 0.112±0.016, in 1974; these were significantly lower than levels found in 1972 (Walker, 1976).

The golden eagle, a carnivore that is terminal in its food chain, is surveyed by the US Department of the Interior under the National Pesticides Monitoring Program. During March 1964 and February 1970, 48 of 169 golden eagles from 22 states were found to contain PCBs in either brain, heart, kidney, liver, muscle or fat, in concentrations ranging from less than 1 to 19 µg/g on a wet weight basis (Reidinger & Crabtree, 1974). Measurable PCB residues were found in 102 of 129 woodcocks collected from 23 eastern and midwestern states from October 1970-February 1971 at levels up to 0.43 µg/g on a wet weight basis (Clark & McLane, 1974).

In 1971 the eggs of birds with a variety of feeding habits, collected from the Niagara Peninsula in Ontario, Canada, were analysed for PCB residues: the highest levels were found in the eggs of carnivores which feed in the aquatic food chain. The mean residue concentrations in whole eggs ranged from 0.05-74 µg/g (Frank et al., 1975).

PCBs were found in all samples of fatty tissues from a variety of arctic mammals, including porpoises, seals, foxes, a polar bear and a sheep killed in 1972, at levels of 0.3-21.0 µg/g (Clausen et al., 1974).

(g) Human tissues and secretions

(i) Fat

Biros et al. (1970) reported 200 and 600 µg/g PCBs in samples of human fat. Price & Welch (1972) found 115-180 µg/g in adipose tissue, 250 µg/g in liver and 125 µg/g in kidney (all on fat basis) in one autopsy case.

In a national US survey of human adipose tissue during the years 1973 and 1974, 35.1 and 40.3%, respectively, of the tissues collected contained

1 µg/g PCBs or more (wet weight basis). The compounds most frequently encountered were penta-, hexa- and heptachlorobiphenyls. Of fat samples evaluated in 1972, 34.2% had non-detectable levels, while in 1974 this percentage had decreased to 9.1; the number of samples containing less than 1 µg/g increased from 33.3% to 50.6%, and the number of samples containing from 1-2 µg/g PCBs increased from 27.3% to 35.4% (Kutz & Strassman, 1976; Yobs, 1972).

Grant *et al.* (1976) found that females in Canada have lower PCB residues than males; the majority of Canadians had mean adipose tissue residues of 1-2 µg/g (fat basis).

Acker & Schulte (1974) found a mean value of 6.8-10 µg/g PCBs (fat basis) in human fat in the Federal Republic of Germany; no significant differences were found in samples from various parts of the country. In a similar investigation of human perirenal fat in Austria, Pesendorf *et al.* (1973) found a mean value of 3.5 µg/g (fat basis).

Hattula *et al.* (1976) analysed PCB levels in 73 autopsy samples from Finland. The highest level (14.2 µg/g adipose tissue on a wet weight basis) was found in a sample from a fisherman. Mean values on a wet weight basis were 1.97 µg/g for adipose tissue, 0.15 µg/g for liver and 0.12 µg/g for brain; the equivalent values on a fat basis were 2.8, 2.6 and 1.19 µg/g, respectively.

In Japan, levels of PCBs in human fat (fat basis) ranged from 0.61-18.04 µg/g, with means ranging from 1.48-7.5 µg/g; the highest values were found in fishermen. The PCBs were mostly those with higher chlorine contents (Tatsukawa, 1976).

(ii) Milk

Risebrough & Brodine (1970) reported that the mean PCB level in human milk in two California cities was about 0.06 µg/ml of whole milk. Assuming a daily intake of 150 g/kg bw, breast-fed infants in these cities ingested approximately 9 µg/kg/day PCBs. Analysis of the lipid fraction of 80 samples of human milk from various areas of the US showed that all except 2 had concentrations ranging from 0.4-10.6 µg/g. The average concentration in all samples was approximately 1.7 µg/g (New York State Health Planning Commission, 1977).

In Ontario, Canada, mean PCB levels in human milk were approximately 1 µg/g lipid (Grant *et al.*, 1976).

Mean levels in human milk in the Federal Republic of Germany were 3.5 µg/g of lipid fraction, or 0.1 µg/ml of whole milk (Acker & Schulte, 1970).

An evaluation of the levels of PCBs in human milk throughout Japan in 1972, 1973 and 1974 showed that one-fourth of all samples in 1974 contained more than 0.033 µg/ml; this level represents the tolerance limit in babies, on the basis of an 'acceptable daily intake' of 5 µg/kg bw and a daily milk intake of 150 ml/kg bw (Tatsukawa, 1976).

(iii) Blood

Forthy-three percent of 723 plasma samples from persons in South Carolina not occupationally exposed to pesticides were found to contain PCBs, with a maximum of 0.029 µg/ml (Finklea et al., 1972).

2.3 Analysis

Methods of analysis for PCBs in environmental samples have been reviewed (Hutzinger et al., 1974a; Safe, 1976).

The Association of Official Analytical Chemists has published an Official First Action for the determination of PCBs in poultry fat, fish, paper and paperboard, including procedures for extraction, clean-up (using Florisil chromatography) and analysis (by gas chromatography with electron capture detection) (Horwitz, 1975).

Column chromatographic separation of PCBs from interfering organochlorine pesticides has been studied using (a) silicic acid (Armour & Burke, 1970; Huckins et al., 1976; McNeil et al., 1977); (b) Florisil (magnesium silicate) (Slade, 1975; US Food and Drug Administration, 1977b); and (c) alumina (American National Standards Institute, Inc., 1974). Additional methods for separating PCBs from other organic compounds include thin-layer chromatography, paper chromatography, solvent partition, sulphuric acid and alcoholic potassium hydroxide treatment (Hutzinger et al., 1974a), ion-exchange resins (Coburn et al., 1977) and gel permeation chromatography (Johnson et al., 1976).

The US PCB producer uses gas chromatography and electron capture detection to determine the amount and types of PCBs in environmental samples of air, water and sediment; a level of detection of 2 µg/kg and an absolute sensitivity of 0.5 ng have been reported (American National Standards Institute, Inc., 1974).

Water samples and bottom sediments from 46 sites in the US were analysed quantitatively for PCBs using a method involving solvent extraction and removal of chlorinated pesticides using alumina and silica gel column chromatography. Basic identification of PCBs was made by gas chromatography with electron capture detection; confirmation was by gas chromatography-mass spectrometry when sample size and concentrations were sufficient. The lower level of detection was 0.1 µg/l of water and 5.0 µg/kg of bottom sediment (Crump-Wiesner et al., 1974).

Gas chromatography-mass spectrometry procedures have been developed to determine mg/kg levels of PCBs in breast milk, fat and environmental samples (Hutzinger et al., 1974a). Better separation, obtained by use of capillary glass columns instead of packed columns, allows improved quantification of individual PCB isomers (Falkmer et al., 1978; Mattsson & Nygren, 1976; Rappe et al., 1977).

In a quantitative thin-layer chromatography method for the analysis of PCBs in tissue, eggs and fat, the spots are developed using silver nitrate and ultra-violet irradiation. The limit of detection was 0.5 mg/kg, with a precision of ±0.05 mg/kg (Bush & Lo, 1973).

Seventeen mono-, di-, tri- and tetrachlorobiphenyls were analysed by separation on a cross-linked polystyrene-type cation exchange resin and identified by ultra-violet absorption. The method is applicable to traces of these chlorinated biphenyls in water (Hanai & Walton, 1977).

A technique utilizing charcoal column chromatography has been used to separate PCB isomers according to the number of chlorine atoms in the *ortho* positions (adjacent to the biphenyl bridge) (Jensen & Sundström, 1974a). Chromatography using an alumina column has also been used for the separation of polychlorinated dibenzofurans from PCBs (Rappe et al., 1977).

High-pressure liquid chromatography and detection using ultra-violet absorption have been used to separate 47 'pure' PCB isomers and 7 commercial mixtures. Preliminary results indicate that the level of detection is below 1 mg/kg and may approach 100 µg/kg (Brinkman et al., 1976).

3. Biological Data Relevant to the Evaluation of Carcinogenic Risk to Humans

3.1 Carcinogenicity and related studies in animals[1]

(a) Oral administration

Mouse: Groups of 12 8-week-old male dd mice were administered Kanechlor 300, 400 or 500 at concentrations of 100, 250 or 500 mg/kg of diet for 32 weeks; 6 control mice received the basal diet. After 32 weeks. 7/12 mice given 500 mg/kg Kanechlor 500 had liver nodules; 5 of these were heptocellular carcinomas. No metastases or tumours were seen in other organs, and no tumours occurred in the other groups (Ito et al., 1973).

Two groups of 50 male 5-6-week-old BALB/cJ mice were fed 300 mg Aroclor 1254/kg of diet either for 6 months followed by the control diet for 5 months or for 11 months; 2 additional groups of 50 mice were fed the basal diet. At the end of 11 months, 1/24 surviving mice treated for 6 months with Aroclor 1254 had a hepatoma, while 9/22 mice that received the treatment for 11 months had 10 hepatomas. In addition, liver lesions described

---

[1]The Working Group was aware of several studies, either in progress or completed but unpublished, to assess the carcinogenicity of Aroclor 1254 and Kanechlor 500 by oral administration to rats and of Kanechlor 300 and 400 by oral administration to mice and rats (IARC, 1978b).

as adenofibrosis were observed in all 22 mice fed Aroclor 1254 continuously for 11 months. No hepatomas were seen in controls (Kimbrough & Linder, 1974).

Rat: Groups of 10 male and 10 female 3-4-week-old Sherman rats were fed 0, 20, 100, 500 or 1000 mg Aroclor 1260/kg diet or 0, 20, 100 or 500 mg Aroclor 1254/kg of diet for 8 months. Several animals given the two highest dose levels died before 6 months. In animals fed Aroclor 1260, lesions described as adenofibrosis of the liver occurred in 2 males fed 1000 mg/kg and in 1, 1 and 4 females fed 100, 500 and 1000 mg/kg, respectively. A higher incidence of this lesion occurred in rats fed Aroclor 1254, i.e., in 10 males and 9 females fed 500 mg/kg and in 1 male and 7 females fed 100 mg/kg (Kimbrough *et al.*, 1972).

A group of 10 male and 10 female 10-week-old Donryu rats were fed 38.5-616 mg Kanechlor 400/kg of diet for 400 days; a control group of 5 males and 5 females received a basal diet. Multiple adenomatous nodules of the liver were observed in 6/10 treated females; these lesions did not occur in male rats nor in the controls (Kimura & Baba, 1973).

Groups of 30 8-week-old male Wistar rats were fed Kanechlor 300, 400 or 500 at levels of 0, 100, 500 or 1000 mg/kg of diet. The Kanechlors produced cholangiofibrosis in 2/15, 2/10 and 4/13 rats, respectively, when fed at the 1000 mg/kg level, while lower doses were ineffective. Hepatic nodular hyperplasia occurred with all 3 compounds, the incidence increasing with degree of chlorination and with concentration in the diet: 100 mg/kg Kanechlor 300 produced hyperplasia in 1/22 animals; 100 mg/kg Kanechlor 400 in 2/16, and 1000 mg/kg Kanechlor 400 in 3/10; and with Kanechlor 500, 100 mg/kg produced hyperplasia in 3/25, 500 mg/kg in 5/16 and 1000 mg/kg in 5/13 animals (Ito *et al.*, 1974).

A group of 200 3-4-week-old female Sherman rats were fed a diet containing 100 mg Aroclor 1260/kg of diet for 21-22 months; 200 female rats served as controls. The authors calculated that the PCB intake declined from 11.6 mg/kg bw/day during the first week of exposure to 6.1 mg/kg bw/day at 3 months of exposure and to 4.3 mg/kg bw/day at 20 months. Twenty-six of 184 exposed rats had hepatocellular carcinomas, and 144/184 had liver lesions described as neoplastic nodules. One hepatocellular carcinoma and no neoplastic nodules were observed among 173 control rats. The incidence of extrahepatic tumours did not differ between the treated and control rats (Kimbrough *et al.*, 1975).

3.2 Other relevant biological data

(a) Experimental systems

(i) Toxic effects

LD_{50}'s for the various Aroclors are summarized in Table 8. More detailed reports on the acute toxic effects of PCBs are given by Allen &

Norback (1976), Allen et al. (1974a, 1975), Aulerich et al. (1973), Harris & Rose (1972), McCune et al. (1962), Platonow & Karstad (1973), Vos (1972) and Vos & Koeman (1970).

Rats are more resistant to the toxic effect of PCBs than are primates (see Table 9). The major changes in rats fed Aroclor 1248, 1254 or 1262 for 6 weeks were liver hypertrophy, marked fatty infiltration and degeneration of parenchymal cells. Mixtures containing isomers with a lower chlorine content were more toxic (Allen & Abrahamson, 1973). In animals fed these mixtures for 1 year, an increase in total serum lipids and cholesterol, a transient increase in serum triglycerides (with Aroclor 1254), liver hypertrophy and focal areas of hepatocellular degeneration occurred (Allen et al., 1976). Rats fed Aroclor 1254 or 1260 for 8 months also showed liver-cell hypertrophy, with cytoplasmic inclusions and brown pigment in the Kupffer cells (Kimbrough et al., 1972).

The major pathological changes that occur in chickens exposed to PCB mixtures and isomers include subcutaneous oedema, ascites, hydropericardium and loss of fat. There was usually marked involution of the thymus, atrophy of the spleen, increased liver weight, congestion, mild necrosis and marked fatty infiltration of the liver as well as widespread haemorrhage and focal necrosis in the kidney (Harris & Rose, 1972; McCune et al., 1962; Platonow et al., 1973). The decreasing order of acute toxicity for various hexachlorobiphenyls (HCBs) in one-day-old cockerels was 3,3',4,4',5,5'-HCB > 2,2',4,4',6,6'-HCB > 2,2',3,3',4,4'-HCB, 2,2',4,4',5,5'-HCB and 2,2',3,3',-6,6'-HCB (Biocca et al., 1976).

In rabbits painted with a mixture of PCB isomers containing approximately 42% chlorine on alternate days, 9-15 times, lesions included fatty degeneration and necrosis of the liver, thinning of the prickle-cell layer and hyperkeratosis of the skin (Miller, 1944). Wedel et al. (1943) painted the dorsal skin of rabbits daily with a mixture of Halowax and 0.3, 0.6 or 0.9 g Aroclor. The higher dose produced death; the intermediate dose caused mottling, fatty degeneration and necrosis of the liver. The skin was reddened and small papules and blisters formed on the surface; there was eventually desquamation of the epidermal layers. Skin application of Phenoclor DP-6, Clophen A-60 or Aroclor 1260 produced loss of weight, hyperplasia and hyperkeratosis of the epidermal and follicular epithelium; centrilobular degeneration, centrilobular liver-cell atrophy, focal necrosis and cytoplasmic hyaline degeneration; hydropic degeneration of the convoluted tubules and tubular dilatation with cast formation; and atrophy of the thymus, lymphopenia and elevated excretion of faecal coproporphyrin and protoporphyrin (Vos & Beems, 1971). The single isomer, 2,2',4,4',5,5'-HCB produced more severe liver lesions but less severe skin lesions in rabbits than did the mixture Aroclor 1260 (Vos & Notenboom-Ram, 1972).

Oral administration of 300 mg Aroclor 1242 or 1254 to rabbits produced liver hypertrophy; in animals treated with Aroclor 1254 there was also atrophy of the uterus (Koller & Zinkl, 1973).

Table 8

LD$_{50}$ (mg/kg bw) of Aroclor mixtures

Species (route)	1221	1232	1242	1248	1254	1260	1262
Mouse (oral)[a]							
Rat (oral)[b]	4000[c]	4500[c]	8700[c]	11,000[c]	2000	10,000[d]	11,300[d]
Adult rat (oral)[e]					4000-10,000	4000-10,000	
Weanling rat (oral)[f]					1300	1300	
Adult rat (i.v.)[f]					350		
30-Day-old rat (oral)[g]					1300-1400		
120-Day-old rat (oral)[g]					2000-2500		
Rabbit (skin)[b] (MLD)	2000-3200[c]	1300-2000[c]	800-1300[c]	800-1300[c]		1300-2000[d]	1300-3200[d]

[a] From Tanaka et al. (1969)
[b] From Panel on Hazardous Trace Substances (1972)
[c] Administration of undiluted compound
[d] Administration of 50% solution in corn oil
[e] From Kimbrough (1974)
[f] From Linder et al. (1974)
[g] From Grant & Phillips (1974)

Table 9

Responses of primates and rats to PCBs

Response	Man[a]	Monkey[b]	Rat[c]
Susceptibility to toxicity	High	High	Moderate
Acne	Yes	Yes	No
Hyperpigmentation of skin	Yes	Only infants	No
Alopecia	NA	Yes	No
Hyperactive Meibomian glands	Yes	Yes	No
Conjunctivitis	Yes	Yes	No
Oedema of eyelids	Yes	Yes	No
Subcutaneous oedema	Yes	Yes	No
Keratin cysts in hair follicles	Yes	Yes	No
Hyperplasia of hair follicle epithelium	Yes	Yes	No
Gastric hyperplasia	NA	Yes	No
Thymic atrophy	NA	NA	Yes
Hepatic hypertrophy	Yes	Yes	Yes
Liver enzyme change	NA	Yes	Yes
Decreased no. of red-blood cells	Yes	Yes	No
Decreased haemoglobin	Yes	Yes	No
Serum hyperlipidaemia	Yes	Hypolipidaemia	Yes
Leucocytosis	Yes	Yes	No

NA = not available
[a] From Kuratsune (1972)
[b] From Allen (1975); Allen & Norback (1976); Allen et al. (1974a)
[c] From Allen & Abrahamson (1973); Allen et al. (1976)

Monkeys fed Aroclor 1248 at levels ranging from 2.5-300 mg/kg of diet developed acne, swelling of the upper eyelids, loss of eyelashes, alopecia, subcutaneous oedema, hypertrophic hyperplastic gastritis, ulceration, hypoproteinaemia and anaemia (Allen, 1975; Allen & Norback, 1973, 1976; Allen et al., 1973, 1974a; Barsotti et al., 1976) (see also Table 9).

(ii) Immune effects

Generalized effects of PCBs on the lymphoid system of guinea-pigs (Vos & de Roij, 1972; Vos & Van Driel-Grootenhuis, 1972), rabbits (Vos & Beems, 1971) and monkeys (Allen, 1975) include lymphopenia, thymic and splenic atrophy and decrease in the number of circulating lymphocytes; hepatitis was seen in ducks (Friend & Trainer, 1970). These effects may be related to altered immune competence.

In monkeys exposed transplacentally and by milk to PCBs, the lymph nodules of the spleen were extremely small with no apparent germinal centres (Allen & Barsotti, 1976).

(iii) Endocrine effects

S.c. administration of Aroclor 1221, 1232, 1242 and 1248 to rats had an oestrogenic effect (Bitman & Cecil, 1970). Female nonhuman primates fed Aroclor 1248 for 6 months showed an increased level of urinary ketosteroids and a prolongation of their menstrual cycles with increased bleeding (Barsotti & Allen, 1975; Barsotti et al., 1976).

(iv) Embryotoxicity, teratogenicity and reproductive effects

No studies were available in which a single PCB isomer or its hydroxylated metabolite was tested or in which common contaminants of PCB mixtures were investigated.

Male and female Sherman rats were fed on a diet containing 100 mg Aroclor 1254/kg of diet for 67 or 186 days prior to mating: increased mortality of the offspring and reduced mating performance of surviving offspring were observed. When diets containing 500 mg Aroclor 1260/kg of diet were fed to similar groups of rats for 67 or 186 days prior to mating, reduced litter size and decreased survival of the offspring were observed. Following administration of oral doses of 100 mg/kg bw/day Aroclor 1254 to pregnant Sherman rats during days 7-15 of gestation, only 30% of the offspring survived to weaning; no apparent teratogenic effects were seen. Reproduction and pup survival were not affected following oral doses of 50 mg/kg bw/day Aroclor 1254 or 100 mg/kg bw/day Aroclor 1260 administered during days 7-15 of gestation (Linder et al., 1974).

Treatment of pregnant Sprague-Dawley rats throughout gestation with 500 mg Kanechlor 500/kg of diet resulted in decreased maternal weight gain and suppressed food consumption; 20 and 500 mg Kanechlor 300/kg of diet and 500 mg Kanechlor 500/kg of diet caused decreases in foetal weights.

No significant differences in the mean numbers of implants, viable foetuses and resorption rates were observed between controls and treated groups (Shiota, 1976b).

Offspring of Sprague-Dawley rats that received 20 or 100 mg/kg bw/day Kanechlor 500 on days 8-14 or 15-21 of gestation learned to swim in a water-filled multiple T-maze test more slowly than controls, and a dose-response relationship was evident; activity and emotionality on the open field were not affected (Shiota, 1976a).

Aroclors 1221, 1232, 1242, 1248, 1254 and 1268 were fed to chickens at a level of 20 mg/kg of diet: Aroclors 1232, 1242, 1248 and 1254 reduced hatchability and caused abnormalities in the embryos, the most common of which were oedema and unabsorbed yolk (Cecil et al., 1974).

Oral administration of 12.5-50 mg/kg bw/day Aroclor 1254 to rabbits for 28 days of gestation was foetotoxic, as indicated by abortions and still-borns; maternal deaths were also observed. Dead foetuses from treated animals showed no consistent skeletal abnormalities (Villeneuve et al., 1971a).

The infants of female monkeys fed Aroclor 1248 before, during and after pregnancy showed facial acne and oedema, swelling of the eyelids, loss of facial hair, including eyelashes, and hyperpigmentation of the skin within 2 months after birth. At death, hyperplastic gastritis and keratinization of the hair follicles of the eyelashes were seen (Allen & Barsotti, 1976).

When female rhesus monkeys were exposed to 2.5 and 5.0 mg Aroclor 1248/kg of diet for 6 months or longer, early abortions occurred in 4/6 impregnated animals that received 5.0 mg/kg of diet and in 3/8 that had 2.5 mg/kg of diet. The offspring that survived were small, with short long bones, small head circumference and reduced crown-to-rump lengths and had detectable levels of PCBs in their tissues (Barsotti et al., 1976).

(v) Absorption, distribution and excretion

When ^{14}C-penta- and -hexachlorobiphenyls were administered to mice, the rate of faecal excretion of radioactivity was related to the positions of chlorine atoms in the molecules (Gage & Holm, 1976).

Hashimoto et al. (1976) fed ^{14}C-labelled PCBs resembling Kanechlor 400 and 600 to rats over periods of up to one year. After administration of ^{14}C-Kanechlor 400, between 1.9 and 4.9% of the radioactivity was excreted in the urine, the amounts being higher in the groups given PCBs for longer periods. After administration of ^{14}C-Kanechlor 600 only 0.56% of the dose was excreted in the urine; 47-68% of the dose of Kanechlor 400 and 57% of Kanechlor 600 were excreted in the faeces. The level of radioactivity was highest in fat, intermediate in skin, adrenal gland, aorta and sciatic nerve and lowest in blood and other tissues.

Of 2-, 3- and 4-chlorobiphenyls, 4,4'- and 2,2-dichlorobiphenyls, 2,2',5,5'- and 3,3',4,4'-tetrachlorobiphenyls, 2,2',4,5,5'-pentachlorobiphenyl and 2,2',4,4',5,5'-hexachlorobiphenyl fed to rats, less than 10% was excreted in the faeces, indicating a high degree of absorption, metabolism or localization in the tissues (Albro & Fishbein, 1972).

Matthews & Anderson (1975a) studied the distribution and excretion of ^{14}C-labelled 4-chloro-, 4,4'-dichloro-, 2,2',4,5,5'-pentachloro- and 2,2',4,4',5,5'-hexachlorobiphenyls in rats at increasing time intervals of 15 min to 42 days following their i.v. administration. The percentage of the total radioactivity accounted for by metabolites of the PCBs decreased as chlorination of the PCBs increased. Excretion accounted for more than 90% of the mono-, di- and pentachlorobiphenyls, whereas hexachlorobiphenyl was excreted very slowly (less than 20% over 42 days).

In rats given i.v. doses of ^{14}C-2,2',4,5,5'-pentachlorobiphenyl, more than 90% of the total dose was removed from the blood within 10 minutes. Most of the radioactivity was deposited initially in liver and muscle and was later translocated to skin and fat. Metabolism was a prerequisite to excretion, which was primarily *via* the biliary route. Excretion in the urine amounted to less than 7% and ceased after 8-9 days (Matthews & Anderson, 1975b).

In rats, when excretion *via* the bile duct was prevented by ligation, a small portion of intraperitoneally or intravenously injected 4-chloro-, 4,4'-dichloro-, 2,2',5,5'- and 2,3',4,4'-tetrachloro- and hexachlorobiphenyls was excreted unchanged in the faeces, indicating that they can pass directly into the gastrointestinal tract (Hutzinger *et al.*, 1972b; Yoshimura & Yamamoto, 1975).

Following i.v. injection to rats of 4-chloro-, 4,4'-dichloro- and 2,2',4,4',5,5'-hexachlorobiphenyls, these compounds were found in hair, suggesting that this may serve as a route of PCB excretion (Matthews *et al.*, 1976).

In Wistar rats administered ^{3}H-Kanechlor 400 in olive oil by skin painting on the clipped dorsal skin, radioautography demonstrated the presence of PCBs in the cells surrounding the hair follicles and, to a lesser extent, in peripheral nerve fibres, capillary endothelial cells in the dermis and in the cytoplasm of hepatic parenchymal cells. No activity was demonstrated in cells of the small intestine (Nishizumi, 1976a).

Over 90% of a single oral dose (1.5 or 3.0 g/kg bw) of Aroclor 1248 was absorbed from the gastrointestinal tract of monkeys as determined by chromatographic analysis of excreta; the major route was *via* biliary excretion into the gastrointestinal tract. By the 14th day after PCB administration, 5.6% of the original dose had been eliminated in the urine and faeces (Allen *et al.*, 1974b).

(vi) Metabolism

The identified metabolites of various PCB isomers are listed in Table 10. The identified and hypothetical pathways of some PCBs are shown in Figure 2. The major identified metabolites of PCBs are phenols and their glucuronides. The identification of *trans*-dihydrodiols in the excreta of rodents and nonhuman primates suggests the formation of an electrophilic precursor, an arene oxide, as a common metabolic intermediate for the PCBs (Gardner et al., 1973; Hsu et al., 1975; Norback et al., 1976). Such a pathway is further suggested by the isolation of methylsulphone derivatives of PCBs (Mio et al., 1976), whose formation can be explained by the reaction of the intermediary arene oxide with methionyl residues in proteins and subsequent cleavage (Yoshimura & Yoshihara, 1976) (Fig. 3).

PCB administration has been found to result in increased synthesis, hepatic content and excretion of porphyrins in rats, quails and chickens; and this has been associated with an increase in liver mitochondrial δ-aminolevulinic acid synthetase (Goldstein et al., 1975, 1976; Grote et al., 1975; Vos & Koeman, 1970).

PCBs are widely used as enzyme inducers in research laboratories throughout the world. Exposure of rats, rabbits, monkeys, chicks and rainbow trout to PCBs resulted in increased activity of hepatic microsomal mixed-function oxidases and some other enzymes, including aniline hydroxylase, *O*- and *N*-demethylases, esterase, aryl hydrocarbon hydroxylase, NADPH, cytochrome *c* reductase, *para*-nitrophenol glucuronyl transferase, uridine diphosphoglucuronosyltransferase and nitroreductase, or a significant increase in the level of cytochrome P-450 (Allen & Abrahamson, 1972; Allen et al., 1973; Alvares & Kappas, 1975; Chen & DuBois, 1973; Fujita et al., 1971; Goldstein et al., 1975, 1976; Iverson et al., 1975; Johnstone et al., 1974; Lidman et al., 1976; Litterst & van Loon, 1974; Litterst et al., 1972; Turner & Green, 1974; Vainio, 1974; Villeneuve et al., 1971b; Wolff & Hesse, 1977). Glucose-6-phosphatase levels in rat liver decreased (Allen & Abrahamson, 1973; Litterst et al., 1972). There are indications that the chlorine content of the PCB mixtures is related to the level of increased enzymatic activity: mixtures containing a lower percentage of chlorines were less active than those containing a higher percentage (Allen & Abrahamson, 1973; Bickers et al., 1972; Chen & DuBois, 1973; Goldstein et al., 1975).

(vii) Modifying factors

The effect of PCBs on the carcinogenic activity of a variety of chemicals has been studied. Oral administration of Kanechlor 400 after (but not before or simultaneously with) the administration of 3'-methyl-4-dimethyl-aminoazobenzene induced a higher incidence of hepatomas in rats (Kimura et al., 1976). Subsequent administration of Kanechlor 500 enhanced the incidence of liver tumours produced in rats by *N*-nitrosodiethylamine (Nishizumi, 1976b). Kanechlor 500 inhibited the hepatocarcinogenicity of

Table 10

Metabolism of PCB isomers

Species	Parent compound	Products	References
Mouse	2,2',5,5'-tetrachloro-biphenyl	3- and 4-methylsulphonyl-2,2',5,5'-tetrachlorobiphenyl; 3- and 4-methylthio-2,2',5,5'-tetrachlorobiphenyl	Mio et al. (1976)
Mouse	2,2',4,4',5,5'-hexachlorobiphenyl	2,2',4,4',5,5'-hexachloro-3-biphenylol	Jensen & Sundström (1974b)
Rat	4-chlorobiphenyl	monohydroxylated derivative; dihydroxylated derivative	Hutzinger et al. (1972b)
Rat	2,3-dichlorobiphenyl	2,3-dichloro-4'-biphenylol	Goto et al. (1974a)
Rat	2,4,6-trichlorobiphenyl	4'-hydroxylated derivative; 3',4'-dihydroxylated derivative;	Goto et al. (1974a)
Rat (liver microsomes)	2,2',5-trichlorobiphenyl	monohydroxylated derivative	Ghiasuddin et al. (1976)
Rat	2,2',5,5'-tetrachloro-biphenyl	3-monohydroxylated derivative; 2,2',5,5'-tetrachloro-3-biphenylol; trans-3,4-dihydro-2,2',5,5'-tetrachloro-3,4-biphenyldiol	Van Miller et al. (1975); Norback et al. (1976)
Rat	3,3',4,4'-tetrachloro-biphenyl	2- and 5-hydroxylated derivatives; 3- and 5-hydroxylated derivatives	Yoshimura & Yamamoto (1974); Yamamoto & Yoshimura (1973)
Rat (liver microsomes)	2,2',5,5'-tetrachloro-biphenyl	monohydroxylated derivative	Ghiasuddin et al. (1976)
Rat	2,3,4,5,6-pentachloro-biphenyl	3'- and 4'-hydroxylated derivatives; 3',4'-dihydroxylated derivative	Goto et al. (1974b)

Table 10 - Metabolism of PCB isomers (continued)

Species	Parent compound	Products	References
Rat	2,2',4,5,5'-pentachloro-biphenyl	2,2',4,5,5'-pentachloro-3'-biphenylol; 3',4'-dihydroxy-2,2',4,5,5'-pentachloro-3',4'-biphenyldiol; 2,2',4,5,5'-pentachloro-3',4'-biphenyldiol	Chen et al. (1976) Goto et al. (1975)
Rat	2,2',4,4',6-pentachloro-biphenyl 2,2',4,6,6'-pentachloro-biphenyl 2,3',4,4',6-pentachloro-biphenyl 2,3',4,5',6-pentachloro-biphenyl	monohydroxylated derivative monohydroxylated derivative monohydroxylated derivative monohydroxylated derivative	
Rat	2,3,3',4,4'-pentachloro-biphenyl	no hydroxylated products	Yamamoto et al. (1976)
Rat (liver microsomes)	2,2'4,5,5'-pentachloro-biphenyl	monohydroxylated derivative	Ghiasuddin et al. (1976)
Rat	2,2',4,4',5,5'-hexa-chlorobiphenyl	no hydroxylated products	Hutzinger et al. (1972b)
Rat	2,2',4,4',6,6'-hexa-chlorobiphenyl	monohydroxylated derivative	Goto et al. (1975)
Rat	2,2',4,4',5,5'-hexa-chlorobiphenyl	2,2',4,4',5,5'-hexachloro-3-biphenylol	Jensen & Sundström (1974b)
Rabbit	4-chlorobiphenyl	4-chloro-4'-biphenylol	Block & Cornish (1959)

Table 10 — Metabolism of PCB isomers (continued)

Species	Parent compound	Products	References
Rabbit	4'-chlorobiphenyl	4'-chloro-4-biphenylol; 4'-chloro-3,4-biphenyldiol; 4'-chloro-3-methoxy-4-biphenylol; 4'-chloro-4-methoxy-3-biphenylol	Safe et al. (1975a)
Rabbit (liver microsomes)	4'-chlorobiphenyl	4'-chloro-4-biphenylol; 4'-chloro-3,4-biphenyldiol	Wyndham et al. (1976)
Rabbit	4,4'-dichlorobiphenyl	4,4'-dichloro-3-biphenylol; 3,4'-dichloro-4-biphenylol; 4'-chloro-4-biphenylol	Safe et al. (1976)
Rabbit	2,2',5,5'-tetrachlorobiphenyl	2,2',5,5'-tetrachloro-3-biphenylol; 2,2',5,5'-tetrachloro-4-biphenylol; $trans$-3,4-dihydro-2,2',5,5'-tetrachloro-3,4-biphenyldiol	Gardner et al. (1973)
Rabbit	2,2',4,4',5,5'-hexachlorobiphenyl	monohydroxylated derivative; monohydroxy with chlorine shift; monohydroxymethoxylated derivative	Hutzinger et al. (1974b)
Pigeon	4-chlorobiphenyl	monohydroxylated derivative	Hutzinger et al. (1972b)
Pigeon	4,4'-dichlorobiphenyl	monohydroxylated derivative	Hutzinger et al. (1972b)
Pigeon	2,2',5,5'-tetrachlorobiphenyl	monohydroxylated derivative	Hutzinger et al. (1972b)
Pigeon	2,2',4,4',5,5'-hexachlorobiphenyl	no hydroxylated products	Hutzinger et al. (1972b)
Chicken	2,2',4,4',5,5'-hexachlorobiphenyl	$meta$-hydroxylated derivative; $penta$-chlorobiphenyl derivative; $penta$-chlorotrihydroxylated derivative	McKinney (1976)

Table 10 - Metabolism of PCB isomers (continued)

Species	Parent compound	Products	References
Trout	4-chlorobiphenyl	no hydroxylated products	Hutzinger et al. (1972b)
Trout	4,4-dichlorobiphenyl	no hydroxylated products	Hutzinger et al. (1972b)
Trout	2,2',5,5'-tetrachlorobiphenyl	no hydroxylated products; conjugated metabolites; 4-hydroxy-2,2',5,5'-tetrachlorobiphenyl	Hutzinger et al. (1972b); Melancon & Lech (1976)
Trout	2,2',4,4',5,5'-hexachlorobiphenyl	no hydroxylated metabolites	Hutzinger et al. (1972b)
Goat	4'-chlorobiphenyl	4'-chloro-4-biphenylol; 4'-chloro-3,4-biphenyldiol	Safe et al. (1975b)
Goat	4,4'-dichlorobiphenyl	4,4'-dichloro-3-biphenylol	Safe et al. (1975b)
Cow	4'-chlorobiphenyl	4'-chloro-4-biphenylol	Safe et al. (1975b)
Rhesus monkey	2,2',5,5'-tetrachlorobiphenyl	monohydroxylated derivatives; dihydroxylated derivatives; trans-3,4-dihydro-3,4-tetrachlorobiphenyldiol; hydroxy-3,4-dihydro-3,4-tetrachlorobiphenyldiol	Hsu et al. (1975)

Figure 2[a]

Metabolic pathways for 2,2',5,5'-tetrachlorobiphenyl

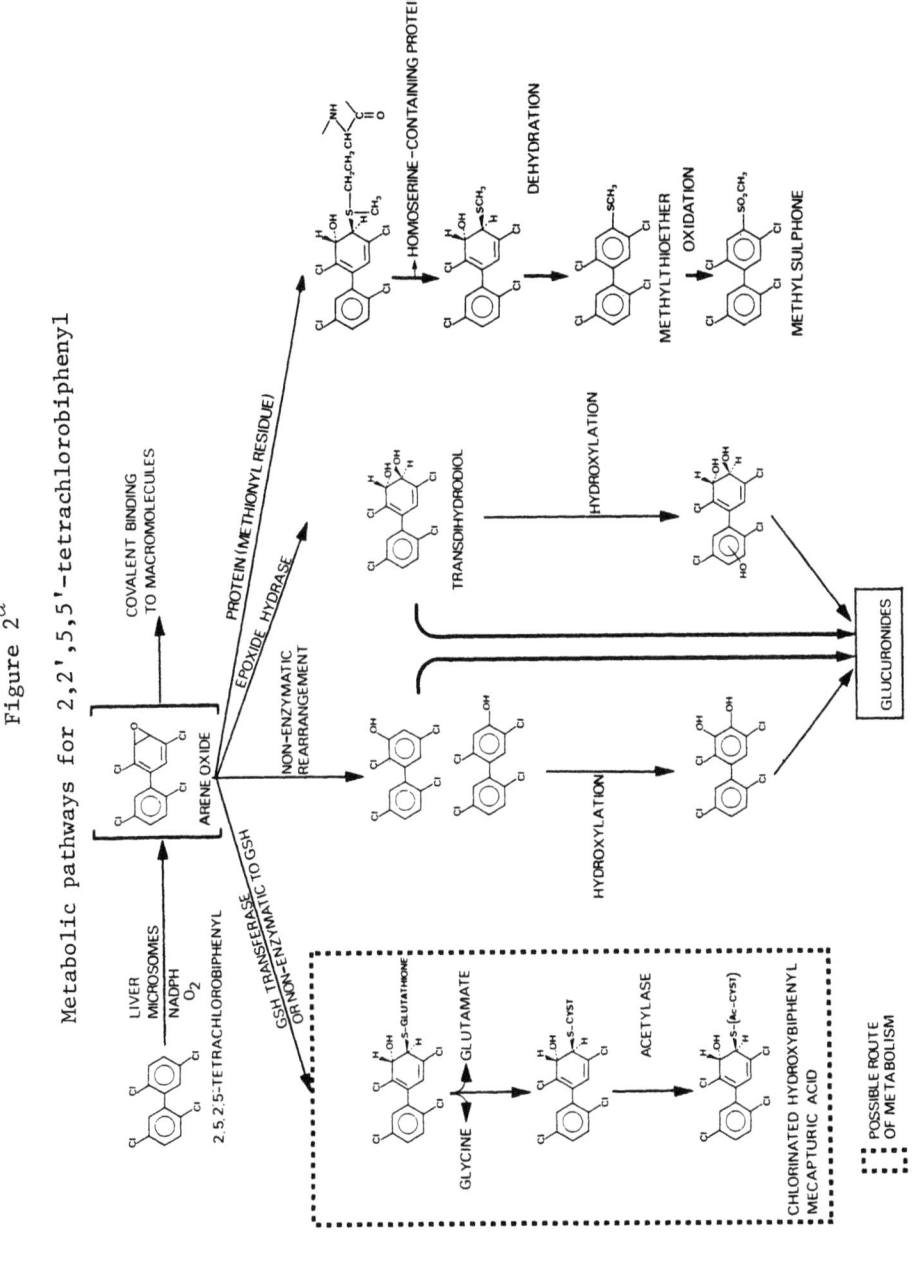

[a] From Allen & Norback, 1977; Yoshimura & Yoshihara (1976)

3'-methyl-4-dimethylaminoazobenze, *N*-2-fluorenylacetamide and *N*-nitrosodiethylamine when given orally with these compounds to rats (Makiura *et al.*, 1974). Kanechlor 400 and 500 enhanced the hepatocarcinogenicity of α-BHC in mice (Nagasaki *et al.*, 1975).

(viii) Mutagenicity

Two PCB mixtures (Aroclor 1221 and Aroclor 1268), 2,2',5,5'-tetrachlorobiphenyl and 4-chlorobiphenyl were tested in *Salmonella typhimurium* TA1538 in the presence of a microsomal activation system from rabbit liver. Only 4-chlorobiphenyl and Aroclor 1221 were significantly mutagenic (Wyndham *et al.*, 1976).

Aroclor 1242 when given as a single oral dose of 1250, 2500 or 5000 mg/kg bw or as 4 daily doses of 500 mg/kg bw/day did not produce significant chromosomal abnormalities in rat bone marrow or spermatogonia. Similarly, 5 oral doses of 75, 150 or 300 mg/kg bw/day Aroclor 1254 did not produce significant chromosomal abnormalities in rat bone marrow (Green *et al.*, 1973, 1975a).

Single oral doses of 625, 1250 or 2500 mg/kg bw or 5 daily doses of 125 or 250 mg/kg bw Aroclor 1242 did not increase the number of dead implantations in rats. Aroclor 1254, when given to rats by oral intubation in 5 daily doses of 75, 150 or 300 mg/kg bw or when fed at a dietary level of 25 or 100 mg/kg of diet for 70 days, did not cause an increase in dominant lethal mutations (Green *et al.*, 1975b).

Hoopingarner *et al.* (1972) observed no chromosomal aberrations in human lymphocyte cultures exposed to 100 mg/kg Aroclor 1254.

Clophen 30 and Clophen 50 had no genetic effect in *Drosophila melanogaster*; tests involved determination of loss of sex chromosomes and of non-disjunction (Nilsson & Ramel, 1974).

Peakall *et al.* (1972) have reported that embryos from second generation ring doves fed 10 mg Aroclor 1254/kg of diet exhibited chromosomal aberrations.

(b) Humans

(i) Toxic effects

A comparison of the toxic effects of PCBs in humans and animals is given in Table 9.

Between March 1968 and April 1975, 1291 cases of 'Yusho' (literally, 'oil disease') were reported in the western part of Japan in persons found or presumed to have ingested rice oil contaminated with Kanechlor 400 (Higuchi, 1976; Kuratsune *et al.*, 1976). In the initial (1968) outbreak, samples of oil were found to contain 800-1000 mg/kg PCB and some up to 3000 mg/kg (Kuratsune, 1976), and 5 mg/kg polychlorinated dibenzofurans (PCDFs) (Kuratsune et al., 1976). Thus, the PCB that contaminated the rice contained about 250 times more PCDFs than do normal PCBs. More than ten different PCDFs were found, the most abundant of which was the 2,3,7,8-tetrachlorinated compound (Rappe et al., 1977); however, tissue samples from Yusho patients contained predominantly penta- and hexachloro derivatives of unknown structure (Nagayama et al., 1977). The average total intake of Kanechlor 400 in a group of 146 patients who used the most highly contaminated oil was estimated to be 2.0 g (10 mg PCDF), and the minimal amount of Kanechlor 400 ingested was 0.5 g (Kuratsune, 1976).

The most common clinical features in the early phase of Yusho included ocular discharge and swelling of the upper eyelids, dermatological symptoms, mainly associated with follicular keratosis (Urabe & Koda, 1976), and various nervous and general symptoms (Kuratsune, 1976). In addition, the following were described: hypertriglyceridaemia with normal plasma cholesterol, adrenocortical and ovarian dysfunction, bronchitis (cough and phlegm), hepatomegaly, hypobilirubinaemia, low immunoglobulin levels and (possibly) disturbed calcium metabolism (Hirayama, 1976) (see Table 9). The most marked dermatological and ophthalmic manifestations gradually improved over the years following exposure (Kuratsune et al., 1976).

By May 1974, 24 Yusho patients had died. Five of the dead patients were autopsied between 12 and 46 months after the first appearance of skin lesions. A stillborn baby was also autopsied 4 months after the appearance of skin lesions in the mother. Histopathological changes likely to be due to PCB ingestion included hyperkeratosis of the hair follicles and increased melanin pigment in the basal layer of the epidermis. In addition, 3 of 5 autopsied patients had proliferation of the duct epithelium of the oesophageal glands (Kikuchi & Masuda, 1976). Detectable levels of PCBs were present, particular in fatty tissue (Kuratsune, 1976).

Kuroki & Masuda (1977) found pronounced differences in the levels of various PCBs in Yusho patients and in normal persons. In the blood of Yusho patients, the PCB found most abundantly is the 2,2',3,4,4',5-hexachlorobiphenyl, and that in normal persons 2,3',4,4',5-pentachlorobiphenyl.

Nagayama et al. (1977) analysed the liver of a man thought to have ingested 560 mg Kanechlor 400 and 2.8 mg PCDFs. When this patient died, approximately 44 months after use of the toxic rice oil had been discontinued, it was found that 0.22% of the ingested PCDFs but only 0.011% of the ingested PCBs remained in the liver.

Industrial exposure to PCBs and/or chlorinated naphthalenes has resulted in chloracne (Drinker et al., 1937; Greenburg et al., 1939; Jones & Alden, 1936; Meigs et al., 1954; Schwartz, 1936, 1943). Three fatal cases were reported among seven workers with jaundice who were exposed to PCBs and/or chlorinated naphthalenes; one patient also had chloracne, acute yellow liver atrophy and cirrhosis, and one other also had acute yellow liver atrophy (Drinker et al., 1937).

The highest levels of PCBs in 73 Finnish autopsy cases studied were found in the liver or adipose tissue of three persons who had died of cancer: 1 of lung cancer, 1 of leukaemia and 1 of liver cancer (Hattula et al., 1976).

(ii) Exposure *in utero* and during lactation

A study was made of the infants of 11 mothers affected by Yusho and of 2 unaffected wives of patients; 2 of the Yusho mothers had stillbirths. The average amount of oil ingested by the mothers during pregnancy ranged from 0.3-2.6 litres. Ten of the babies had greyish or dark-brown pigmentation of the skin, and 5 had similar pigmentation of the gingiva and/or nails; increased ocular discharge was present in 9; and 12 of the 13 infants were small when compared with the national average (Funatsu et al., 1972; Kikuchi et al., 1969; Kuratsune, 1976; Taki et al., 1969). Histological examinations of one stillborn baby and of four of the small babies showed hyperkeratosis and atrophy of the epidermis and cystic dilatation of the hair follicles, particularly of the head. There was also a marked increase in melanin pigment in the basal cells of the epidermis. Detectable but low levels of PCBs (0.1-1.8 mg/kg on fat basis) were present in the mesenteric fat, skin and liver (Kikuchi & Masuda, 1976; Kikuchi et al., 1969). Babies born to patients even 3 years after severe PCB exposure tended to show pigmentation of the skin on the back and the gingiva, although the degree of pigmentation was less than that of babies born to the same mothers up to one year after the poisoning (Kuratsune, 1976).

Mothers' milk contaminated with PCBs also appears to be a source of exposure for infants: one baby showed signs of Yusho even though the mother had ingested the contaminated rice oil only after the baby was delivered. The infant began to show signs after 3-4 months of breast feeding (Kuratsune, 1976; Yoshimura, 1974).

Developmental abnormalities have also been observed in PCB-intoxicated infants. Premature eruption of teeth was observed in 2 cases, and larger frontal and occipital fontanelles, exophthalmos and the maintenance of an abnormally wide sagittal suture were observed in 3 others. No other obvious malformations were found (Funatsu et al., 1972).

3.3 Case reports and epidemiological studies[1]

Deaths that occurred up to 5½ years after first exposure to PCBs (September, 1973) among 1200 Yusho patients in Japan [see section 3.2 (b)] were reported by Kuratsune (1976) and Urabe (1974). Nine (41%) of 22 deaths were due to malignant neoplasms. Three of the tumours occurred in

---

[1] The Working Group was aware of two retrospective mortality studies in progress of a cohort of workers exposed occupationally to PCBs (tumours at all sites) (IARC, 1978c).

the stomach, one in the liver (with cirrhosis), two in the lungs and one in the breast, and two were malignant lymphomas. An additional liver cancer was mentioned in connection with one of the stomach cancers, but it is not clear whether this was an additional primary cancer or a metastasis from the stomach. No estimate was given from population rates of the numbers or types of tumours expected in this group. In 1975, 29 deaths were reported among 1291 Yusho patients, but causes of death were not given (Kuratsune, 1976; Omae, 1975).

In another report (Bahn *et al.*, 1976, 1977), two malignant melanomas were diagnosed in 31 workers heavily exposed to Aroclor 1254 and probably to other chemicals [It was estimated from the data of the Third National Cancer Survey (National Cancer Institute, 1975) that 0.04 malignant melanomas would have been expected in this group of individuals]. Among 41 other workers also, but less heavily exposed to Aroclor 1254, one additional melanoma was diagnosed. Among the 31 heavily exposed workers, three other individuals developed 4 cancers at other sites, including 2 in the pancreas.

4. Summary of Data Reported and Evaluation[1]

4.1 Experimental data

Five polychlorinated biphenyl mixtures have been tested in mice and/or rats only by oral administration. Kanechlor 500 and Aroclor 1254 are carcinogenic in mice, and Aroclor 1260 is carcinogenic in rats; all induced benign and malignant liver-cell tumours. In an experiment in rats of only one year's duration, Kanechlor 500, 400 and 300 induced liver lesions described as multiple hyperplastic nodules.

---

[1]Subsequent to the finalization of this monograph by the Working Group in October 1977, the Secretariat became aware of a study carried out under the NCI Bioassay Programme (NCI, 1978). Groups of 24 male and 24 female Fischer 344 rats were given Aroclor 1254 at concentrations of 25, 50 or 100 mg/kg of diet for 104-105 weeks, when surviving animals were killed. No statistically significant differences between tumour incidences in experimental and control animals were seen. However, a few carcinomas and adenocarcinomas of the gastrointestinal tract were observed in treated animals; no such tumours occurred in controls. Hepatocellular hyperplastic nodules were observed in 11/48, 17/46 and 29/48 treated animals, compared with none in controls.

4.2 Human data

Human exposure to small amounts of polychlorinated biphenyls is widespread as a result of environmental contamination and the high stability of these compounds. They are commonly found in human tissues. Unusually high levels of exposure to polychlorinated biphenyls have occurred among workers manufacturing or using them and in Japanese who consumed rice oil accidentally contaminated with Kanechlor 400. The latter showed acute and chronic toxic effects.

An apparent excess of malignant melanoma has been reported in workers exposed to Aroclor 1254. No melanomas were reported in 9 persons who died from cancer among the 1200 Japanese heavily exposed to Kanechlor 400, but these deaths all occurred within $5\frac{1}{2}$ years of first exposure. Neither the workers exposed occupationally nor the Japanese were exposed solely to polychlorinated biphenyls.

4.3 Evaluation

There is experimental evidence of a carcinogenic effect of some polychlorinated biphenyls in rodents. The epidemiological data provide suggestive evidence of a relationship between exposure to polychlorinated biphenyls and the development of malignant melanoma. Efforts should be made to obtain both confirmatory experimental and epidemiological evidence; in particular, continuing follow-up of survivors of the Yusho episode is necessary. In the meantime, for practical purposes, polychlorinated biphenyls should be regarded as if they were carcinogenic to humans.

Almost without exception, polychlorinated biphenyls contain various levels of polychlorinated dibenzofurans as contaminants, and the polychlorinated biphenyls responsible for the Yusho episode in Japan were found to contain an unusually high level of polychlorinated dibenzofurans. It is not known if and to what extent polychlorinated dibenzofurans play a role in the observed carcinogenic effects of polychlorinated biphenyls.

5. References

Acker, L. & Schulte, E. (1970) Über das Vorkommen von chlorierten Biphenylen und Hexachlorobenzol neben chlorierten Insektiziden in Humanmilch und menschlichem Fettgewebe. Naturwissenschaften, 57, 497

Acker, L. & Schulte, E. (1974) Chlorkohlenwasserstoff im menschlichen Fett. Naturwissenschaften, 61, 32

Albro, P.W. & Fishbein, L. (1972) Intestinal absorption of polychlorinated biphenyls in rats. Bull. environ. Contam. Toxicol., 8, 26-31

Allen, J.R. (1975) Response of the nonhuman primate to polychlorinated biphenyl exposure. Fed. Proc., 34, 1675-1679

Allen, J.R. & Abrahamson, L.J. (1972) Enzymatic changes in the liver of rats fed chlorinated biphenyls, triphenyls and DDT. Proc. Am. Chem. Soc., 12, 97-111

Allen, J.R. & Abrahamson, L.J. (1973) Morphological and biochemical changes in the liver of rats fed polychlorinated biphenyls. Arch. environ. Contam. Toxicol., 1, 265-280

Allen, J.R. & Barsotti, D.A. (1976) The effects of transplacental and mammary movement of PCBs on infant rhesus monkeys. Toxicology, 6, 331-340

Allen, J.R. & Norback, D.H. (1973) Polychlorinated biphenyl- and triphenyl-induced gastric mucosal hyperplasia in primates. Science, 179, 498-499

Allen, J.R. & Norback, D.H. (1976) Pathological responses of primates to polychlorinated biphenyl exposure. In: Proceedings of the National Conference on Polychlorinated Biphenyls, Chicago, 1975, EPA-560/6-75-004, Washington DC, Environmental Protection Agency, pp. 43-49

Allen, J.R. & Norback, D.H. (1977) Carcinogenic potential of the polychlorinated biphenyls. In: Hiatt, H.H., Watson, J.D. & Winsten, J.A., eds, Origins of Human Cancer, Book A, Incidence of Cancer in Humans, Cold Spring Harbor, N.Y., Cold Spring Harbor Laboratory, pp. 173-186

Allen, J.R., Abrahamson, L.J. & Norback, D.H. (1973) Biological effects of polychlorinated biphenyls and triphenyls on the subhuman primate. Environ. Res., 6, 344-354

Allen, J.R., Carstens, L.A. & Barsotti, D.A. (1974a) Residual effects of short-term, low-level exposure of nonhuman primates to polychlorinated biphenyls. Toxicol. appl. Pharmacol., 30, 440-451

Allen, J.R., Norback, D.H. & Hsu, I.C. (1974b) Tissue modifications in monkeys as related to absorption, distribution, and excretion of polychlorinated biphenyls. Arch. environ. Contam. Toxicol., 2, 86-95

Allen, J.R., Carstens, L.A., Abrahamson, L.J. & Marlar, R.J. (1975) Responses of rats and nonhuman primates to 2,5,2',5'-tetrachlorobiphenyl. Environ. Res., 9, 265-273

Allen, J.R., Carstens, L.A. & Abrahamson, L.J. (1976) Responses of rats exposed to polychlorinated biphenyls for fifty-two weeks. I. Comparison of tissue levels of PCB and biological changes. Arch. environ. Contam. Toxicol., 4, 404-419

Alvares, A.P. & Kappas, A. (1975) Induction of aryl hydrocarbon hydroxylase by polychlorinated biphenyls in the foeto-placental unit and neonatal livers during lactation. FEBS Lett., 50, 172-174

Alvares, A.P., Bickers, D.R. & Kappas, A. (1974) Induction of drug-metabolizing enzymes and aryl hydrocarbon hydroxylase by microscope immersion oil. Life Sci., 14, 853-860

American National Standards Institute, Inc. (1974) American National Standard Guidelines for Handling and Disposal of Capacitor- and Transformer-grade Askarels Containing Polychlorinated Biphenyls, ANSI C107.1-1974, New York, pp. 20-35

Anon. (1977) EPA bans discharge of PCBs into waterways. Chem. Week, 120, 40

Armour, J.A. & Burke, J.A. (1970) Method for separating polychlorinated biphenyls from DDT and its analogs. J. Assoc. off. anal. Chem., 53, 761-768

Aulerich, R.J., Ringer, R.K. & Iwamoto, S. (1973) Reproductive failure and mortality in mink fed on Great Lakes fish. J. Reprod. Fertil. Suppl., 19, 365-376

Bahn, A.K., Rosenwaike, I., Herrmann, N., Grover, P., Stellman, J. & O'Leary, K. (1976) Melanoma after exposure to PCB's. New Engl. J. Med., 295, 450

Bahn, A.K., Grover, P., Rosenwaike, I., O'Leary, K. & Stellman, J. (1977) PCB? and melanoma. New Engl. J. Med., 296, 108

Barsotti, D.A. & Allen, J.R. (1975) Effects of polychlorinated biphenyls on reproduction in the primate (Abstract No. 675). Fed. Proc., 34, 338

Barsotti, D.A., Marlar, R.J. & Allen, J.R. (1976) Reproductive dysfunction in rhesus monkeys exposed to low levels of polychlorinated biphenyls (Aroclor 1248). Food Cosmet. Toxicol., 14, 99-103

Bennett, H.S. & Albro, P.W. (1973) PCB's in microscopic immersion oil. Science, 181, 990

Bickers, D.R., Harber, L.C., Kappas, A. & Alvares, A.P. (1972) Polychlorinated biphenyls: comparative effects of high and low chlorine containing Aroclors(R) on hepatic mixed function oxidase. Res. Commun. chem. Pathol. Pharmacol., 3, 505-512

Biocca, M., Moore, J.A., Gupta, B.N. & McKinney, J.D. (1976) Toxicology of selected symmetrical hexachlorobiphenyl isomers. I. Biological responses in chicks and mice. In: Proceedings of the National Conference on Polychlorinated Biphenyls, Chicago, 1975, EPA-560/6-75-004, Washington DC, Environmental Protection Agency, pp. 67-72

Biros, F.J., Walker, A.C. & Medbery, A. (1970) Polychlorinated biphenyls in human adipose tissue. Bull. environ. Contam. Toxicol., 5, 317-323

Bitman, J. & Cecil, H.C. (1970) Estrogenic activity of DDT analogs and polychlorinated biphenyls. J. agric. Food Chem., 18, 1108-1112

Block, W.D. & Cornish, H.H. (1959) Metabolism of biphenyl and 4-chlorobiphenyl in the rabbit. J. biol. Chem., 234, 3301-3302

Bowes, G.W., Mulvihill, M.J., Simoneit, B.R.T., Burlingame, A.L. & Risebrough, R.W. (1975) Identification of chlorinated dibenzofurans in American polychlorinated biphenyls. Nature (Lond.), 256, 305-307

Brinkman, U.A.T., Seetz, J.W.F.L. & Reymer, H.G.M. (1976) High-speed liquid chromatography of polychlorinated biphenyls and related compounds. J. Chromatogr., 116, 353-363

Buser, H.-R. (1978) Polychlorinated Dibenzo-p-Dioxins and Dibenzofurans: Formation, Occurrence and Analysis of Environmental Hazardous Compounds, Thesis, Department of Organic Chemistry, University of Umeå, Umeå, Sweden

Bush, B. & Lo, F.-C. (1973) Thin-layer chromatography for quantitative polychlorinated biphenyl analysis. J. Chromatogr., 77, 377-388

Carey, A.E. & Gowen, J.A. (1976) PCB's in agricultural and urban soil. In: Proceedings of the National Conference on Polychlorinated Biphenyls, Chicago, 1975, EPA-560/6-75-004, Washington DC, Environmental Protection Agency, pp. 195-198

Cecil, H.C., Bitman, J., Lillie, R.J., Fries, G.F. & Verrett, J. (1974) Embryotoxic and teratogenic effects in unhatched fertile eggs from hens fed polychlorinated biphenyls (PCBs). Bull. environ. Contam. Toxicol., 11, 489-495

Chen, P.R., McKinney, J.D. & Matthews, H.B. (1976) Metabolism of 2,4,5,2',5'-pentachlorobiphenyl in the rat. Drug Metab. Disp., 4, 362-367

Chen, T.S. & DuBois, K.P. (1973) Studies on the enzyme inducing effect of polychlorinated biphenyls. Toxicol. appl. Pharmacol., 26, 504-512

Clark, D.R., Jr & McLane, M.A.R. (1974) Chlorinated hydrocarbon and mercury residues in woodcock in the United States, 1970-71. Pestic. Monit. J., 8, 15-22

Clausen, J., Braestrup, L. & Berg, O. (1974) The content of polychlorinated hydrocarbons in arctic mammals. Bull. environ. Contam. Toxicol., 12, 529-534

Coburn, J.A., Valdmanis, I.A. & Chau, A.S.Y. (1977) Evaluation of XAD-2 for multiresidue extraction of organochlorine pesticides and polychlorinated biphenyls from national waters. J. Assoc. off. anal. Chem., 60, 224-228

Crump-Wiesner, H.J., Feltz, H.R. & Yates, M.L. (1974) A study of the distribution of polychlorinated biphenyls in the aquatic environment. Pestic. Monit. J., 8, 157-161

Dennis, D.S. (1976) Polychlorinated biphenyls in the surface waters and bottom sediments of the major drainage basins of the United States. In: Proceedings of the National Conference on Polychlorinated Biphenyls, Chicago, 1975, EPA-560/6-75-004, Washington DC, Environmental Protection Agency, pp. 183-194

Drinker, C.K., Warren, M.F. & Bennett, G.A. (1937) The problem of possible systemic effects from certain chlorinated hydrocarbons. J. ind. Hyg. Toxicol., 19, 283-311

Durfee, R.L. (1976) Production and usage of PCB's in the United States. In: Proceedings of the National Conference on Polychlorinated Biphenyls, Chicago, 1975, EPA-560/6-75-004, Washington DC, Environmental Protection Agency, pp. 103-107

Eichner, M. (1976) Über Rückstandsbestimmungen von chlorierten Insecticiden und polychlorierten Biphenylen in Fischen des Bodensees, des Oberrheins und dessen Zuflüssen sowie in diesen Gewässern. II. Z. Lebensmittelunters., 161, 327-336

Elder, D. (1976) PCBs in NW Mediterranean coastal waters. Mar. Pollut. Bull., 7, 63-64 [Chem. Abstr., 85, 105075m]

Elkins, H.G.B. (1959) The Chemistry of Industrial Toxicology, 2nd ed., New York, John Wiley and Sons, p. 153

Falkmer, S., Marklund, S., Mattsson, P.E. & Rappe, C. (1978) Hepatomas and other neoplasms in a primitive vertebrate, the Atlantic hagfish (*Myxine glutinosa*). A histopathological and chemical study with particular reference to the role of polychlorinated biphenyls, some chlorinated pesticides, and aflatoxins. Ann. N.Y. Acad. Sci. (in press)

Finklea, J., Priester, L.E., Creason, J.P., Hauser, T., Hinners, T. & Hammer, D.I. (1972) Polychlorinated biphenyl residues in human plasma expose a major urban pollution problem. Am. J. Public Health, 62, 645-651

Frank, R., van Hove Holdrinet, M. & Rapley, W.A. (1975) Residue of organochlorine compounds and mercury in birds' eggs from the Niagara Peninsula, Ontario. Arch. environ. Contam. Toxicol., 3, 205-218

Friend, M. & Trainer, D.O. (1970) Polychlorinated biphenyl: interaction with duck hepatitis virus. Science, 170, 1314-1316

Fujita, S., Tsuji, H., Kato, K., Saeki, S. & Tsukamoto, H. (1971) Effect of biphenyl chlorides on rat liver microsomes. Fukuoka acta med., 62, 30-34

Fujiwara, K. (1975) Environmental and food contamination with PCB's in Japan. Sci. Total Environ., 4, 219-247

Fukushima, M. (1974) M.S. Thesis, College of Agriculture, Ehime University, Japan

Funatsu, I., Yamashita, F., Ito, Y., Tsugawa, S., Funatsu, T., Yoshikane, T., Hayashi, M., Kato, T., Yakushiji, M., Okamoto, G., Yamasaki, S., Arima, T., Kuno, T., Ide, H. & Ide, I. (1972) Polychlorobiphenyls (PCB) induced fetopathy. I. Clinical observation (Abstract No. 72-2360). Kurume med. J., 19, 43-51

Gage, J.C. & Holm, S. (1976) The influence of molecular structure on the retention and excretion of polychlorinated biphenyls by the mouse. Toxicol. appl. Pharmacol., 36, 555-560

Gardner, A.M., Chen, J.T., Roach, J.A.G. & Ragelis, E.P. (1973) Polychlorinated biphenyls: hydroxylated urinary metabolites of 2,5,2',5'-tetrachlorobiphenyl identified in rabbits. Biochem. biophys. Res. Commun., 55, 1377-1384

Ghiasuddin, S.M., Menzer, R.E. & Nelson, J.O. (1976) Metabolism of 2,5,2'-trichloro-, 2,5,2',5'-tetrachloro-, and 2,4,5,2',5'-pentachlorobiphenyl in rat hepatic microsomal system. Toxicol. appl. Pharmacol., 36, 187-194

Goldstein, J.A., Hickman, P., Burse, V.W. & Bergman, H. (1975) A comparative study of two polychlorinated biphenyl mixtures (Aroclors 1242 and 1016) containing 42% chlorine on induction of hepatic porphyria and drug metabolizing enzymes. Toxicol. appl. Pharmacol., 32, 461-473

Goldstein, J.A., McKinney, J.D., Lucier, G.W., Hickman, P., Bergman, H. & Moore, J.A. (1976) Toxicological assessment of hexachlorobiphenyl isomers and 2,3,7,8-tetrachlorodibenzofuran in chicks. II. Effects on drug metabolism and porphyrin accumulation. Toxicol. appl. Pharmacol., 36, 81-92

Goto, M., Sugiura, K., Hattori, M., Miyagawa, T. & Okamura, M. (1974a) Metabolism of 2,3-dichlorobiphenyl-^{14}C and 2,4,6-trichlorobiphenyl-^{14}C in the rat. Chemosphere, 5, 227-232

Goto, M., Sugiura, K., Hattori, M., Miyagawa, T. & Okamura, M. (1974b) Metabolism of 2,3,5,6-tetrachlorobiphenyl-^{14}C and 2,3,4,5,6-pentachlorobiphenyl-^{14}C in the rat. Chemosphere, 5, 233-238

Goto, M., Hattori, M. & Sugiura, K. (1975) Metabolism of pentachloro- and hexachlorobiphenyls in the rat. Chemosphere, 3, 177-180

Graham, J.M. (1976) Levels of PCB's in Canadian commercial fish species. In: Proceedings of the National Conference on Polychlorinated Biphenyls, Chicago, 1975, EPA-560/6-75-004, Washington DC, Environmental Protection Agency, pp. 155-160

Grant, D.L. & Phillips, W.E.J. (1974) The effect of age and sex on the toxicity of Aroclor$^{(R)}$ 1254, a polychlorinated biphenyl, in the rat. Bull. environ. Contam. Toxicol., 12, 145-152

Grant, D.L., Mes, J. & Frank, R. (1976) PCB residues in human adipose tissue and milk. In: Proceedings of the National Conference on Polychlorinated Biphenyls, Chicago, 1975, EPA-560/6-75-004, Washington DC, Environmental Protection Agency, pp. 144-146

Green, S., Palmer, K.A. & Oswald, E.J. (1973) Cytogenetic effects of the polychlorinated biphenyls (Aroclor 1242) on rat bone marrow and spermatogonial cells (Abstract No. 113). Toxicol. appl. Pharmacol., 25, 482

Green, S., Carr, J.V., Palmer, K.A. & Oswald, E.J. (1975a) Lack of cytogenetic effects in bone marrow and spermatogonial cells in rats treated with polychlorinated biphenyls (Aroclors 1242 and 1254). Bull. environ. Contam. Toxicol., 13, 14-22

Green, S., Sauro, F.M. & Friedman, L. (1975b) Lack of dominant lethality in rats treated with polychlorinated biphenyls (Aroclors 1242 and 1254). Food Cosmet. Toxicol., 13, 507-510

Greenburg, L., Mayers, M.R. & Smith, A.R. (1939) The systemic effects resulting from exposure to certain chlorinated hydrocarbons. J. ind. Hyg. Toxicol., 21, 29-38

Griefs (1867) Erzetzung des Wasserstoffs durch Stickstoff in organischen Verbindungen. Tetrazodiphenyl-Verbindungen. J. Prakt. Chem., 101, 91-94

Grote, W., Schmoldt, A. & Benthe, H.F. (1975) Hepatic porphyrin synthesis in rats after pretreatment with polychlorinated biphenyls (PCBs). Acta pharmacol. toxicol., 36, 215-224

Hanai, T. & Walton, H.F. (1977) Chromatography of chlorinated biphenyls on an ion-exchange resin. Anal. Chem., 49, 764-766

Harris, J.R. & Rose, L. (1972) Toxicity of polychlorinated biphenyls in poultry. J. Amer. vet. med. Assoc., 161, 1584-1586

Harvey, G.R., Steinhauer, W.G. & Miklas, H.P. (1974) Decline of PCB concentrations in North Atlantic surface water. Nature (Lond.), 252, 387-388

Hasegawa, H., Sato, M. & Tsuruta, H. (1973) An investigation on the toxicity of some new substances used as PCB replacement, concentrations in air of SAS, KMC-oil and PCBs in carbonless paper producing plants and health examination of workers. Special Research Report 141-211, Research Coordination Bureau, Science and Technology Agency, Tokyo, Japan

Hashimoto, K., Akasaka, S., Takagi, Y., Kataoka, M., Otake, T., Murata, Y., Aburada, S., Kitaura, T. & Uda, H. (1976) Distribution and excretion of [^{14}C]polychlorinated biphenyls after their prolonged administration to male rats. Toxicol. appl. Pharmacol., 37, 415-423

Hattula, M.L., Ikkala, J., Isomäki, M., Määttä, K. & Arstila, A.U. (1976) Chlorinated hydrocarbon residues (PCB and DDT) in human liver, adipose tissue and brain in Finland. Acta pharmacol. toxicol., 39, 545-554

Higuchi, K. (1976) Outline. In: Higuchi, K., ed., PCB Poisoning and Pollution, Tokyo, Kodansha Ltd, pp. 3-7

Hirayama, C. (1976) Clinical aspects of PCB poisoning. In: Higuchi, K., ed., PCB Poisoning and Pollution, Tokyo, Kodansha Ltd, pp. 87-104

Hoopingarner, R., Samuel, A. & Krause, D. (1972) Polychlorinated biphenyl interactions with tissue culture cells. Environ. Health Perspect., 1, 155-158

Horwitz, W., ed. (1975) Official Methods of Analysis of the Association of Official Analytical Chemists, 12th ed., Washington DC, Association of Official Analytical Chemists, pp. 518-528

Hsu, I.C., Van Miller, J.P., Seymour, J.L. & Allen, J.R. (1975) Urinary metabolites of 2,5,2',5'-tetrachlorobiphenyl in the nonhuman primate (38999). Proc. Soc. exp. Biol. (N.Y.), 150, 185-188

Hubbard, H.L. (1964) Chlorinated biphenyl and related compounds. In: Kirk, R.E. & Othmer, D.F., eds, Encyclopedia of Chemical Technology, 2nd ed., Vol. 5, New York, John Wiley and Sons, pp. 289-297

Huckins, J.N., Stalling, D.L. & Johnson, J.L. (1976) Silicic acid chromatographic separation of polychlorinated biphenyls and pesticides: some contaminants and limitations. J. Assoc. off. anal. Chem., 59, 975-981

Hutzinger, O., Safe, S. & Zitko, V. (1972a) Photochemical degradation of chlorobiphenyls (PCBs) Environ. Health Perspect., 1, 15-20

Hutzinger, O., Nash, D.M., Safe, S., DeFreitas, A.S.W., Norstrom, R.J., Wildish, D.J. & Zitko, V. (1972b) Polychlorinated biphenyls: metabolic behavior of pure isomers in pigeons, rats, and brook trout. Science, 178, 312-314

Hutzinger, O., Safe, S. & Zitko, V. (1974a) The Chemistry of PCB's, Cleveland, Ohio, Chemical Rubber Co., pp. 1-2, 41-70, 189-193

Hutzinger, O., Jamieson, W.D., Safe, S., Paulmann, L. & Ammon, R. (1974b) Identification of metabolic dechlorination of highly chlorinated biphenyl in rabbit. Nature (Lond.), 252, 698-699

IARC (1974) IARC Monographs on the Evaluation of Carcinogenic Risk of Chemicals to Man, 7, Some Anti-Thyroid and Related Substances, Nitrofurans and Industrial Chemicals, Lyon, pp. 261-289

IARC (1978a) Long-term hazards of polychlorinated dibenzodioxins and polychlorinated dibenzofurans. IARC intern. tech. Rep. No. 78/001

IARC (1978b) IARC Information Bulletin on the Survey of Chemicals Being Tested for Carcinogenicity, No. 7, Lyon, pp. 89, 124-125, 251, 382

IARC (1978c) Directory of On-Going Research in Cancer Epidemiology, 1978, Lyon (IARC Scientific Publications No. 26), pp. 286 (Abstract No. 754), 350 (Abstract No. 930)

Interdepartmental Task Force on PCBs (1972) PCBs and the Environment, COM-72-10419, Springfield, Virginia, National Technical Information Service, pp. 5, 11

Ito, N., Nagasaki, H., Arai, M., Makiura, S., Sugihara, S. & Hirao, K. (1973) Histopathologic studies on liver tumorigenesis induced in mice by technical polychlorinated biphenyls and its promoting effect on liver tumors induced by benzene hexachloride. J. natl Cancer Inst., 51, 1637-1646

Ito, N., Nagasaki, H., Makiura, S. & Arai, M. (1974) Histopathological studies on liver tumorigenesis in rats treated with polychlorinated biphenyls. Gann, 65, 545-549

Iverson, F., Villeneuve, D.C., Grant, D.L. & Hatina, G.V. (1975) Effect of Aroclor 1016 and 1242 on selected enzyme systems in the rat. Bull. environ. Contam. Toxicol., 13, 456-463

Jelinek, C.F. & Corneliussen, P.E. (1976) Levels of PCB's in the US food supply. In: Proceedings of the National Conference on Polychlorinated Biphenyls, Chicago, 1975, EPA-560/6-75-004, Washington DC, Environmental Protection Agency, pp. 147-154

Jensen, S. & Sundström, G. (1974a) Structures and levels of most chlorobiphenyls in two technical PCB products and in human adipose tissue. Ambio, 3, 70-76

Jensen, S. & Sundström, G. (1974b) Metabolic hydroxylation of a chlorobiphenyl containing only isolated unsubstituted positions - 2,2',4,4',5,5'-hexachlorobiphenyl. Nature (Lond.), 251, 219-220

Johnson, L.D., Waltz, R.H., Ussary, J.P. & Kaiser, F.E. (1976) Automated gel permeation chromatographic cleanup of animal and plant extracts for pesticide residue determination. J. Assoc. off. anal. Chem., 59, 174-187

Johnstone, G.J., Ecobichon, D.J. & Hutzinger, O. (1974) The influence of pure polychlorinated biphenyl compounds on hepatic function in the rat. Toxicol. appl. Pharmacol., 28, 66-81

Jones, J.W. & Alden, H.S. (1936) An acneform dermatergosis. Arch. Dermatol. Syphilol., 33, 1022-1034

Jordan, D. (1977) The town dilemma. Environment, 19, 6-15

Karppanen, E. & Kolho, L. (1973) The concentrations of PCB in human blood and adipose tissue in three different research groups. In: Proceedings of the 2nd PCB Conference, Solna, Sweden, 1972, Stockholm, National Swedish Environmental Protection Board, pp. 124-128

Kikuchi, M. & Masuda, Y. (1976) The pathology of Yusho. In: Higuchi, K., ed., PCB Poisoning and Pollution, Tokyo, Kodansha Ltd, pp. 69-86

Kikuchi, M., Hashimoto, M., Hozumi, M., Koga, K., Oyoshi, S. & Nagakawa, M. (1969) An autopsy case of stillborn of chlorobiphenyls poisoning. Fukuoka acta med., 60, 489-495

Kimbrough, R.D. (1974) The toxicity of polychlorinated polycyclic compounds and related chemicals. CRC Crit. Rev. Toxicol., 2, 445-498

Kimbrough, R.D. & Linder, R.E. (1974) Induction of adenofibrosis and hepatomas of the liver in BALB/cJ mice by polychlorinated biphenyls (Aroclor 1254). J. natl Cancer Inst., 53, 547-552

Kimbrough, R.D., Linder, R.E. & Gaines, T.B. (1972) Morphological changes in livers of rats fed polychlorinated biphenyls. Light microscopy and ultrastructure. Arch. ind. Health, 25, 354-364

Kimbrough, R.D., Squire, R.A., Linder, R.E., Strandberg, J.D., Montali, R.J. & Burse, V.W. (1975) Induction of liver tumors in Sherman strain female rats by polychlorinated biphenyl Aroclor 1260. J. natl Cancer Inst., 55, 1453-1459

Kimura, N.T. & Baba, T. (1973) Neoplastic changes in the rat liver induced by polychlorinated biphenyl. Gann, 64, 105-108

Kimura, N.T., Kanematsu, T. & Baba, T. (1976) Polychlorinated biphenyl(s) as a promoter in experimental hepatocarcinogenesis in rats. Z. Krebsforsch., 87, 257-266

Kleinert, S.J. (1976) Sources of polychlorinated biphenyls in Wisconsin. In: Proceedings of the National Conference on Polychlorinated Biphenyls, Chicago, 1975, EPA-560/6-75-004, Washington DC, Environmental Protection Agency, pp. 124-126

Koller, L.D. & Zinkl, J.G. (1973) Pathology of polychlorinated biphenyls in rabbits. Am. J. Pathol., 70, 363-378

Kornreich, M., Fuller, B., Dorigan, J., Walker, P. & Thomas, L. (1976) Environmental Impact of Polychlorinated Biphenyls, MTR-7006, McLean, Virginia, The MITRE Corporation

Kuratsune, M. (1972) An abstract of results of laboratory examinations of patients with Yusho and of animal experiments. Environ. Health Perspect., 1, 129-136

Kuratsune, M. (1976) Epidemiologic studies on Yusho. In: Higuchi, K., ed., PCB Poisoning and Pollution, Tokyo, Kodansha Ltd, pp. 9-23

Kuratsune, M., Masuda, Y. & Nagayama, J. (1976) Some of the recent findings concerning Yusho. In: Proceedings of the National Conference on Polychlorinated Biphenyls, Chicago, 1975, EPA-560/6-75-004, Washington DC, Environmental Protection Agency, pp. 14-29

Kuroki, H. & Masuda, Y. (1977) Structures and concentrations of the main components of polychlorinated biphenyls retained in patients with Yusho. Chemosphere, 8, 469-474

Kutz, F.W. & Strassman, S.C. (1976) Residues of polychlorinated biphenyls in the general population of the United States. In: Proceedings of the National Conference on Polychlorinated Biphenyls, Chicago, 1975, EPA-560/6-75-004, Washington DC, Environmental Protection Agency, pp. 139-143

Kutz, F.W. & Yang, H.S.C. (1976) A note on polychlorinated biphenyls in air. In: Proceedings of the National Conference on Polychlorinated Biphenyls, Chicago, 1975, EPA-560/6-75-004, Washington DC, Environmental Protection Agency, p. 182

Lidman, U., Förlin, L., Molander, O. & Axelson, G. (1976) Induction of the drug metabolizing system in rainbow trout (*Salmo gairdnerii*) liver by polychlorinated biphenyls (PCBs). Acta pharmacol. toxicol., 39, 262-272

Linder, R.E., Gaines, T.B. & Kimbrough, R.D. (1974) The effect of polychlorinated biphenyls on rat reproduction. Food Cosmet. Toxicol., 12, 63-77

Litterst, C.L. & van Loon, E.J. (1974) Time-course of induction of microsomal enzymes following treatment with polychlorinated biphenyls. Bull. environ. Contam. Toxicol., 11, 206-212

Litterst, C.L., Farber, T.M., Baker, A.M. & van Loon, E.J. (1972) Effect of polychlorinated biphenyls on hepatic microsomal enzymes in the rat. Toxicol. appl. Pharmacol., 23, 112-122

Lloyd, J.W., Moore, R.M., Jr, Woolf, B.S. & Stein, H.P. (1976) Polychlorinated biphenyls. J. occup. Med., 18, 109-113

Longhurst, A.R. & Radford, P.J. (1975) PCB concentrations in North Atlantic surface water. Nature (Lond.), 256, 239

Makiura, S., Aoe, H., Sugihara, S., Hirao, K., Arai, M. & Ito, N. (1974) Inhibitory effect of polychlorinated biphenyls on liver tumorigenesis in rats treated with 3'-methyl-4-dimethylaminoazobenzene, *N*-2-fluorenylacetamide, and diethylnitrosamine. J. natl Cancer Inst., 53, 1253-1257

Matthews, H.B. & Anderson, M.W. (1975a) Effect of chlorination on the distribution and excretion of polychlorinated biphenyls. Drug Metab. Disp., 3, 371-380

Matthews, H.B. & Anderson, M.W. (1975b) The distribution and excretion of 2,4,5,2',5'-pentachlorobiphenyl in the rat. Drug Metab. Disp., 3, 211-219

Matthews, H.B., Domanski, J.J. & Guthrie, F.E. (1976) Hair and its associated lipids as an excretory pathway for chlorinated hydrocarbons. Xenobiotica, 6, 425-429

Mattsson, P.E. & Nygren, S. (1976) Gas chromatographic determination of polychlorinated biphenyls and some chlorinated pesticides in sewage sludge using a glass capillary column. J. Chromatogr., 124, 265-275

McCune, E.L., Savage, J.E. & O'Dell, B.L. (1962) Hydropericardium and ascites in chicks fed a chlorinated hydrocarbon. Poult. Sci., 41, 295-299

McKinney, J.D. (1976) Toxicology of selected symmetrical hexachlorobiphenyl isomers: correlating biological effects with chemical structure. In: Proceedings of the National Conference on Polychlorinated Biphenyls, Chicago, 1975, EPA-560/6-75-004, Washington DC, Environmental Protection Agency, pp. 73-76

McKinney, J.D., Chae, K., Gupta, B.N., Moore, J.A. & Goldstein, J.A. (1976) Toxicological assessment of hexachlorobiphenyl isomers and 2,3,7,8-tetrachlorodibenzofuran in chicks. I. Relationship of chemical parameters. Toxicol. appl. Pharmacol., 36, 65-80

McNeil, E.E., Otson, R., Miles, W.F. & Rajabalee, F.J.M. (1977) Determination of chlorinated pesticides in potable water. J. Chromatogr., 132, 277-286

Meigs, J.W., Albom, J.J. & Kartin, B.L. (1954) Chloracne from an unusual exposure to Arochlor. J. Am. med. Assoc., 154, 1417-1418

Melancon, M.J., Jr & Lech, J.J. (1976) Isolation and identification of a polar metabolite of tetrachlorobiphenyl from bile of rainbow trout exposed to ^{14}C-tetrachlorobiphenyl. Bull. environ. Contam. Toxicol., 15, 181-188

Miller, J.W. (1944) Pathologic changes in animals exposed to a commercial chlorinated diphenyl. Public Health Rep., 59, 1085-1093

Mio, T., Sumino, K. & Mizutani, T. (1976) Sulfur-containing metabolites of 2,5,2',5'-tetrachlorobiphenyl, a major component of commercial PCB's. Chem. Pharm. Bull. (Tokyo), 24, 1958-1960

Monsanto Industrial Chemicals Co. (1975) Product Specifications, Aroclor 1016, Aroclor 1254, St. Louis, Missouri

Morita, M., Nakagawa, J. & Rappe, C. (1978) Polychlorinated dibenzofuran (PCDF) formation from PCB mixture by heat and oxygen. Bull. environ. Contam. Toxicol. (in press)

Murakami, K. & Takeishi, K. (1977) Behavior of heavy metals and PCBs in dredging and treating of bottom deposits. In: Peterson, S.A. & Randolph, K.K., eds, Proceedings of the 2nd US-Japan Experts' Meeting, Management of Bottom Sediments Containing Toxic Substances, Tokyo, 1976, EPA-600/3-77-083, Cowallis, Oregon, US Environmental Protection Agency, pp. 107-126

Nadeau, R.J. & Davis, R.A. (1976) Polychlorinated biphenyls in the Hudson River (Hudson Falls-Ford Edward, New York State). Bull. environ. Contam. Toxicol., 16, 436-444

Nagasaki, H., Tomii, S. & Mega, T. (1975) Factors affecting induction of liver cancer by BHC and PCBs in mice (Abstract No. 235) Jpn. J. Hyg., 30, 134

Nagayama, J., Kuratsune, M. & Masuda, Y. (1976) Department of chlorinated dibenzofurans in Kanechlors and 'Yusho oil'. Bull. environ. Contam. Toxicol., 15, 9-13

Nagayama, J., Masuda, Y. & Kuratsune, M. (1977) Determination of polychlorinated dibenzofurans in tissues of patients with 'Yusho'. Food Cosmet. Toxicol., 15, 195-198

National Cancer Institute (NCI) (1975) Third National Cancer Survey, Bethesda, Md

National Cancer Institute (NCI) (1978) Bioassay of Aroclor$^{(R)}$ 1254 for Possible Carcinogenicity, CAS No. 27323-18-8, NCI-CG-TR-38, DHEW Publication No. (NIH) 78-838, Washington DC, US Department of Health, Education and Welfare

New York State Health Planning Commission (1977) Report of the *ad hoc* Committee on the Health Implications of PCBs in Mothers' Milk, 1977, Albany, New York, Health Advisory Council

Nilsson, B. & Ramel, C. (1974) Genetic tests on *Drosophila melanogaster* with polychlorinated biphenyls (PCB). Hereditas, 77, 319-322

NIOSH (National Institute of Occupational Safety and Health) (1977) Criteria for a Recommended Standard. Occupational Exposure to Polychlorinated Biphenyls (PCBs), DHEW (NIOSH) Publication No. 77-225, Washington DC, US Government Printing Office

Nisbet, I.C.T. (1976a) Criteria Document for PCBs, EPA-440/9-76-021, PB 255 397, Springfield, Virginia, National Technical Information Service

Nisbet, I.C.T. (1976b) Environmental transport and occurrence of PCB's in 1975. In: Proceedings of the National Conference on Polychlorinated Biphenyls, Chicago, 1975, EPA-560/6-75-004, Washington DC, Environmental Protection Agency, pp. 254-256

Nishizumi, M. (1976a) Radioautographic evidence for absorption of polychlorinated biphenyls through the skin. Ind. Health, 14, 41-44

Nishizumi, M. (1976b) Enhancement of diethylnitrosamine hepatocarcinogenesis in rats by exposure to polychlorinated biphenyls or phenobarbital. Cancer Lett., 2, 11-16

Norback, D.H., Seymour, J.L., Knieriem, K.M., Peterson, R.E. & Allen, J.R. (1976) Biliary metabolites of 2,5,2',5'-tetrachlorobiphenyl in the rat. Res. Commun. chem. Pathol. Pharmacol., 14, 527-533

Omae, T. (1975) The fifth report of the study group on 'Yusho'. Fukuoka acta med., 66, 547-558

Organisation for Economic Cooperation and Development (1976) Utilization and Environmental Levels of Certain Chemical Substances. A Case Study Report From Japan, Paris, Environment Directorate, pp. 9-14, 35-40, 42-48

Ouw, H.K., Simpson, G.R. & Siyali, D.S. (1976) Use and health effects of Aroclor 1242, a polychlorinated biphenyl, in an electrical industry. Arch. environ. Health, 31, 189-194

Panel on Hazardous Trace Substances (1972) Polychlorinated biphenyls - environmental impact. Environ. Res., 5, 249-362

Peakall, D.B. & Risebrough, R.W. (1975) PCB's and their environmental effects. CRC Crit. Rev. environ. Control, 5, 469-508

Peakall, D.B., Lincer, J.L. & Bloom, S.E. (1972) Embryonic mortality and chromosomal alterations caused by Aroclor 1254 in ring doves. Environ. Health Perspect., 1, 103-104

Pesendorf, H., Eichler, I. & Glofke, E. (1973) Orientierende Untersuchungen über Rückstände an DDT, anderen Organochlorpestiziden und PCBs in Humanfettproben. Wiener Klin. Wochenschr., 85, 218-222

Platonow, N.S. & Karstad, L.H. (1973) Dietary effects of polychlorinated biphenyls on mink. Can. J. comp. Med., 37, 391-400

Platonow, N.S., Karstad, L.H. & Saschenbrecker, P.W. (1973) Tissue distribution of polychlorinated biphenyls (Aroclor 1254) in cockerels: relation to the duration of exposure and observations on pathology. Can. J. comp. Med., 37, 90-95

Price, H.A. & Welch, R.L. (1972) Occurrence of polychlorinated biphenyls in humans. Environ. Health Perspect., 1, 73-78

Rappe, C., Gară, A., Buser, H.R. & Bosshardt, H.-P. (1977) Analysis of polychlorinated dibenzofurans in Yusho oil using high resolution gas chromatography-mass spectrometry. Chemosphere, 5, 231-236

Reidinger, R.F., Jr & Crabtree, D.G. (1974) Organochlorine residues in golden eagles, United States - March 1964-July 1971. Pestic. Monit. J., 8, 37-43

Risebrough, R. & Brodine, V. (1970) More letters in the wind. Environment, 12, 16-27

Roach, J.A.G. & Pomerantz, I.H. (1974) The finding of chlorinated dibenzofurans in a Japanese polychlorinated biphenyl sample. Bull. environ. Contam. Toxicol., 12, 338-342

Safe, S. (1976) Overview of analytical identification and spectroscopic properties. In: Proceedings of the National Conference on Polychlorinated Biphenyls, Chicago, 1975, EPA-560/6-75-004, Washington DC, Environmental Protection Agency, pp. 94-102

Safe, S. & Hutzinger, O. (1972) The mass spectra of polychlorinated biphenyls. J. chem. Soc., Perkin I, 686-691

Safe, S., Hutzinger, O. & Jones, D. (1975a) The mechanism of chlorobiphenyl metabolism. J. agric. Food Chem., 23, 851-853

Safe, S., Platonow, N. & Hutzinger, O. (1975b) Metabolism of chlorobiphenyls in the goat and cow. J. agric. Food Chem., 23, 259-261

Safe, S., Jones, D. & Hutzinger, O. (1976) Metabolism of 4,4'-dihalogenobiphenyls. J. chem. Soc., Perkin I, 357-359

Schwartz, L. (1936) Dermatitis from synthetic resins and waxes. Am. J. Public Health, 26, 586-592

Schwartz, L. (1943) An outbreak of halowax acne ('cable rash') among electricians. J. Am. med. Assoc., 122, 158-161

Shiota, K. (1976a) Postnatal behavioral effects of prenatal treatment with PCBs (polychlorinated biphenyls) in rats. Okajimas Fol. anat. jpn., 53, 105-114

Shiota, K. (1976b) Embryotoxic effects of polychlorinated biphenyls (Kanechlors 300 and 500) in rats. Okajimas Fol. anat. jpn., 53, 93-104

Slade, P. (1975) IUPAC commission on the development, improvement, and standardization of methods of pesticide residue analysis. J. Assoc. off. anal. Chem., 58, 1244-1255

Staiff, D.C., Quinby, G.E., Spencer, D.L. & Starr, H.G., Jr (1974) Polychlorinated biphenyl emission from fluorescent lamp ballasts. Bull. environ. Contam. Toxicol., 12, 455-463

Subcommittee on the Health Effects of Polychlorinated Biphenyls and Polybrominated Biphenyls (1976) Final Report, July 1976, Washington DC, Department of Health, Education and Welfare

Subcommittee on the Health Effects of Polychlorinated Biphenyls and Polybrominated Biphenyls (1978) Final Report of the Subcommittee of the Health Effects of Polychlorinated Biphenyls and Polybrominated Biphenyls of the DHEW Committee to coordinate toxicology and related programs. Environ. Health Perspect., 24, 129-239

Sundström, G. (1973) Polychlorinated biphenyls. II. Synthesis of some tetra- and pentachlorobiphenyls. Acta chem. scand., 27, 600-604

Taki, I., Hisanaga, S. & Amagase, Y. (1969) Report on Yusho (chlorobiphenyls poisoning): pregnant women and their fetuses. Fukuoka acta med., 60, 471-474

Tanaka, K., Fujita, S., Komatsu, F. & Tamura, N. (1969) Experimental subacute poisoning of chlorobiphenyls, particularly the influence on the serum lipids in rats. Fukuoka acta med., 60, 544-547

Tas, A.C. & Kleipool, R.J.C. (1972) Characterization of the components of technically polychlorinated biphenyl mixtures. II. Bull. environ. Contam. Toxicol., 8, 32-37

Tas, A.C. & de Vos, R.H. (1971) Characterization of four major components in a technical polychlorinated biphenyl mixture. Environ. Sci. Technol., 5, 1216-1218

Tatsukawa, R. (1976) PCB pollution of the Japanese environment. In: Higuchi, K., ed., PCB Poisoning and Pollution, Tokyo, Kodansha Ltd, pp. 147-179

Tatsukawa, R. & Watanabe, I. (1972) Air pollution by PCBs. Shoku No Kagaku, 8, 55-63

Turner, J.C. & Green, R.S. (1974) The effect of a polychlorinated biphenyl (Aroclor$^{(R)}$ 1254) on liver microsomal enzymes in the male rat. Bull. environ. Contam. Toxicol., 12, 669-671

UNEP/WHO (United Nations Environment Programme/World Health Organization (1976) Polychlorinated Biphenyls and Terphenyls, Environ. Health Criteria, 2, Geneva

Urabe, H. (1974) The fourth report of the study group on 'Yusho'. Fukuoka acta med., 65, 1-4

Urabe, H. & Koda, H. (1976) The dermal symptomatology of Yusho. In: Higuchi, K., ed., PCB Poisoning and Pollution, Tokyo, Kodansha Ltd, pp. 105-123

US Environmental Protection Agency (1976) PCBs in the United States. Industrial Use and Environmental Distribution, PB-252 012, Springfield, Virginia, National Technical Information Service, pp. 4-5, 34-35, 54-57, 198-210, 322-334

US Environmental Protection Agency (1977a) Toxic pollutant effluent standards. Fed. Regist., 42, 6532, 6554-6555

US Environmental Protection Agency (1977b) Polychlorinated biphenyls (PCBs). Fed. Regist., 42, 26564-26577

US Food and Drug Administration (1977a) Polychlorinated biphenyls (PCB's). Fed. Regist., 42, 17487-17494

US Food and Drug Administration (1977b) Pesticide Analytical Manual, 3rd ed., Vol. 1, Extraction and Cleanup, Rockville, MD, US Department of Health, Education, and Welfare, sections 117, 201F, 251.01, 251.02, 251.12a, 251.12b, 251.14, 251.15, 251.16, 251.21, 251.22, 251.26

US International Trade Commission (1977) Imports of Benzenoid Chemicals and Products, 1976, USITC Publication 828, Washington DC, US Government Printing Office, p. 24

US Occupational Safety and Health Administration (1977) Air Contaminants. US Code Fed. Regul., Title 29, part 1910.1000, pp. 59-60

Vainio, H. (1974) Enhancement of microsomal drug oxidation and glucuronidation in rat liver by an environmental chemical, polychlorinated biphenyl. Chem.-biol. Interact., 9, 379-387

Van Miller, J.P., Hsu, I.C. & Allen, J.R. (1975) Distribution and metabolism of ^{3}H-2,5,2',5'-tetrachlorobiphenyl in rats. Proc. Soc. exp. Biol. (N.Y.), 148, 682-687

Veith, G.D. & Lee, G.F. (1971) Chlorobiphenyls (PCBs) in the Milwaukee River. Water Res., 5, 1107-1115

Villeneuve, D.C., Grant, D.L., Khera, K., Clegg, D.J., Baer, H. & Phillips, W.E.J. (1971a) The fetotoxicity of a polychlorinated biphenyl mixture (Aroclor$^{(R)}$ 1254) in the rabbit and in the rat. Environ. Physiol., 1, 67-71

Villeneuve, D.C., Grant, D.L., Phillips, W.E.J., Clark, M.L. & Clegg, D.J. (1971b) Effects of PCB administration on microsomal enzyme activity in pregnant rabbits. Bull. environ. Contam. Toxicol., 6, 120-128

Vos, J.G. (1972) Toxicology of PCBs for mammals and for birds. Environ. Health Perspect., 1, 105-117

Vos, J.G. & Beems, R.B. (1971) Dermal toxicity studies of technical polychlorinated biphenyls and fractions thereof in rabbits. Toxicol. appl. Pharmacol., 19, 617-633

Vos, J.G. & Koeman, J.H. (1970) Comparative toxicologic study with polychlorinated biphenyls in chickens with special reference to porphyria, edema formation, liver necrosis, and tissue residues. Toxicol. appl. Pharmacol., 17, 656-668

Vos, J.G. & Notenboom-Ram, E. (1972) Comparative toxicity study of 2,4,5,2',4',5'-hexachlorobiphenyl and a polychlorinated biphenyl mixture in rabbits. Toxicol. appl. Pharmacol., 23, 563-578

Vos, J.G. & de Roij, T. (1972) Immunosuppressive activity of a polychlorinated biphenyl preparation on the humoral immune response in guinea pigs. Toxicol. appl. Pharmacol., 21, 549-555

Vos, J.G. & Van Driel-Grootenhuis, L. (1972) PCB-induced suppression of the humoral and cell-mediated immunity in guinea pigs. Sci. Total Environ., 1, 289-302

Vos, J.G., Koeman, J.H., van der Maas, H.L., ten Noever de Brauw, M.C. & de Vos, R.H. (1970) Identification and toxicological evaluation of chlorinated dibenzofuran and chlorinated naphthalene in two commercial polychlorinated biphenyls. Food Cosmet. Toxicol., 8, 625-633

Walker, C.R. (1976) The occurrence of PCB in the national fish and wildlife monitoring program. In: Proceedings of the National Conference on Polychlorinated Biphenyls, Chicago, 1975, EPA-560/6-75-004, Washington DC, Environmental Protection Agency, pp. 161-176

Webb, R.G. & McCall, A.C. (1972) Identities of polychlorinated biphenyl isomers in Aroclors. J. Assoc. off. anal. Chem., 55, 746-752

Wedel, H. von, Holla, W.A. & Denton, J. (1943) Observations on the toxic effects resulting from exposure to chlorinated naphthalene and chlorinated phenyls with suggestions for prevention. Rubber Age, 53, 419-426

Welti, D. & Sissons, D. (1972) The proton chemical shifts of polychlorinated biphenyls. Org. Magn. Resonance, 4, 309-319

Willford, W.A., Hesselberg, R.J. & Nicholson, L.W. (1976) Trends of polychlorinated biphenyls in three Lake Michigan fishes. In: Proceedings of the National Conference on Polychlorinated Biphenyls, Chicago, 1975, EPA-560/6-75-004, Washington DC, Environmental Protection Agency, pp. 177-181

Wilson, N.K. (1975) Carbon-13 nuclear magnetic resonance. ^{13}C Shieldings and spin-lattice relaxation times in chlorinated biphenyls. J. Am. chem. Soc., 97, 3573-3578

Winell, M.A. (1975) An international comparison of hygienic standards for chemicals in the work environment. Ambio, 4, 34-36

Wolff, T. & Hesse, S. (1977) Species differences of mixed-function oxidase induction between rabbits and rats after pretreatment with polychlorinated biphenyls (PCB's). Biochem. Pharmacol., 26, 783-787

Wyndham, C., Devenish, J. & Safe, S. (1976) The *in vitro* metabolism, macromolecular binding and bacterial mutagenicity of 4-chlorobiphenyl, a model PCB substrate. Res. Commun. chem. Pathol. Pharmacol., 15, 563-570

Yamamoto, H.-A. & Yoshimura, H. (1973) Metabolic studies on polychlorinated biphenyls. III. Complete structure and acute toxicity of the metabolites of 2,4,3',4'-tetrachlorobiphenyl. Chem. pharm. Bull. (Tokyo), 21, 2237-2242

Yamamoto, H.-A., Yoshimura, H., Fujita, M. & Yamamoto, T. (1976) Metabolic and toxicologic evaluation of 2,3,4,3',4'-pentachlorobiphenyl in rats and mice. Chem. pharm. Bull. (Tokyo), 24, 2168-2174

Yobs, A.R. (1972) Levels of polychlorinated biphenyls in adipose tissue of the general population of the nation. Environ. Health Perspect., 1, 79-81

Yoshimura, T. (1974) Epidemiological study on Yusho babies born to mothers who had consumed oil contaminated by PCB. Fukuoka acta med., 65, 74-80

Yoshimura, H. & Yamamoto, H.-A. (1974) Metabolic studies on polychlorinated biphenyls. IV. Biotransformation of 3,4,3',4'-tetrachlorobiphenyl, one of the major components of Kanechlor-400. Fukuoka acta med., 65, 5-11

Yoshimura, H. & Yamamoto, H.-A. (1975) A novel route of excretion of 2,4,3',4'-tetrachlorobiphenyl in rats. Bull. environ. Contam. Toxicol., 13, 681-688

Yoshimura, H. & Yoshihara, S. (1976) Toxicological aspects. II. The metabolic fate of PCB's and their toxicological evaluation. In: Higuchi, K., ed., PCB Poisoning and Pollution, Tokyo, Kodansha Ltd, pp. 41-67

Young, D.R., McDermott, D.J. & Heesen, T.C. (1976) Marine inputs of polychlorinated biphenyls off Southern California. In: Proceedings of the National Conference on Polychlorinated Biphenyls, Chicago, 1975, EPA-560/6-75-004, Washington DC, Environmental Protection Agency, pp. 199-208

Zislavsky, W. (1976) Studies of fish (red eye) in Austrian waters for residues of chlorinated hydrocarbons, PCB's, and mercury during 1973 to 1975. Land-Forstwirtsch. Forsch. Oesterr., 7, 261-274 [Chem. Abstr., 86, 84359k]

POLYBROMINATED BIPHENYLS

POLYBROMINATED BIPHENYLS[1]

A review on polybrominated biphenyls (PBBs) is available (Subcommittee on the Health Effects of Polychlorinated Biphenyls and Polybrominated Biphenyls, 1976, 1978).

1. Chemical and Physical Data

1.1 Synonyms and trade names

Hexabromobiphenyl (pure chemical)

Chem. Abstr. Services Reg. No.: 59080-40-9
Chem. Abstr. Name: 2,2',4,4',5,5'-Hexabromo-1,1'-biphenyl

2,2',4,4',5,5'-Hexabromobiphenyl

Hexabromobiphenyl (technical grade)

Chem. Abstr. Services Reg. No.: 36355-01-8
Chem. Abstr. Name: Hexabromo-1,1'-biphenyl

Hexabromodiphenyl

Firemaster BP-6; Firemaster FF-1[2]

Octabromobiphenyl

Chem. Abstr. Services Reg. No.: 27858-07-7
Chem. Abstr. Name: ar,ar,ar,ar,ar',ar',ar',ar'-Octabromo-1,1'-biphenyl

Octabromodiphenyl

Bromkal 80

Decabromobiphenyl

Chem. Abstr. Services Reg. No.: 13654-09-6
Chem. Abstr. Name: 2,2',3,3',4,4',5,5',6,6'-Decabromo-1,1'-biphenyl

Decabromodiphenyl; perbromobiphenyl

Flammex B-10

---

[1] Considered by the Working Group in Lyon, June 1978
[2] Pulverized form of Firemaster BP-6 containing 2% by weight of an anticaking agent, calcium polysilicate, produced in limited quantities as a developmental product in 1971 and 1972 (Neufeld et al., 1977).

1.2 Structural and molecular formulae and molecular weight

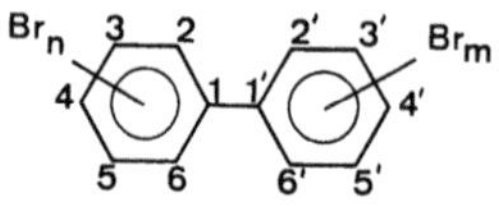

$$\text{Br}_n \underset{5\ 6}{\overset{3\ 2}{\bigcirc}} \underset{6'\ 5'}{\overset{2'\ 3'}{\bigcirc}} \text{Br}_m$$

Hexabromobiphenyl, n + m = 6 $C_{12}H_4Br_6$ Mol. wt: 627.4

Octabromobiphenyl, n + m = 8 $C_{12}H_2Br_8$ Mol. wt: 785.2

Decabromobiphenyl, n + m = 10 $C_{12}Br_{10}$ Mol. wt: 943.1

1.3 Chemical and physical properties of the pure substances

From Neufeld et al. (1977), unless otherwise specified

Hexabromobiphenyl

(a) Description: White solid

(b) Melting-point: 72°C

(c) Spectroscopy data: Ultra-violet spectral data have been tabulated (De Kok et al., 1977).

(d) Solubility: Insoluble in water (11 µg/kg), soluble in acetone (6 g/100 g) and benzene (75 g/100 g); one formulated product is fat soluble (Kay, 1977).

(e) Volatility: Vapour pressure is 0.000076 mm Hg at 90°C.

(f) Stability: Degrades readily in ultra-violet light (Mumma & Wallace, 1975)

(g) Reactivity: Inert; Firemaster FF-1, pyrolysed for 20 min at 380-400°C in open glass tubes and in tubes sealed after nitrogen flushing, gave 2,3,7,8-tetrabromodibenzofuran (O'Keefe, 1978)

Octabromobiphenyl

(a) Description: White solid

(b) Melting-point: 200-250°C

(c) <u>Spectroscopy data</u>: Ultra-violet spectral data have been tabulated (De Kok et al., 1977).

(d) <u>Solubility</u>: Insoluble in water (20-30 μg/kg); slightly soluble in cotton-seed oil (1700 mg/kg), acetone (1.8 g/100 g), and methylene chloride (3.9 g/100 g); soluble in methylene bromide (7.4 g/100 g), benzene (8.10 g/100 g), *ortho*-xylene (10 g/100 g) and chlorobenzene (18.7 g/100 g) (Norris et al., 1973).

(e) <u>Volatility</u>: Weight loss at 250°C is less than 1%

(f) <u>Stability</u>: Degrades readily in ultra-violet light (Mumma & Wallace, 1975)

(g) <u>Reactivity</u>: Inert

Decabromobiphenyl

(a) <u>Description</u>: White solid

(b) <u>Melting-point</u>: 380-386°C

(c) <u>Spectroscopy data</u>: Ultra-violet spectral data have been tabulated (De Kok et al., 1977).

(d) <u>Solubility</u>: Insoluble in water

(e) <u>Volatility</u>: Weight loss at 341°C is less than 5%

(f) <u>Reactivity</u>: Inert

1.4 Technical products and impurities

The bromine content of commercial PBB products ranges from about 77% for hexabromobiphenyl to about 82% for the decabromobiphenyl. Technical grade hexabromobiphenyl (Firemaster BP-6) produced in the US contains 63% hexabromobiphenyl (principally the 2,2',4,4',5,5'-isomer), 14% heptabromobiphenyl (mostly 2,2',3,4,4',5,5'-isomer), 10% pentabromobiphenyl, 2% tetrabromobiphenyl and 11% other bromobiphenyls. Brominated naphthalenes (150 mg/kg pentabromo- and 70 mg/kg hexabromo-) have also been found in Firemaster BP-6 (Hass et al., 1978; Neufeld et al., 1977; Trotter, 1977).

Technical grade octabromobiphenyl produced in the Federal Republic of Germany (Bromkal 80) has the following composition: 72% octabromobiphenyl (consisting of three isomers, present at 42, 16 and 14%), 27% heptabromobiphenyl, 1% hexabromobiphenyl and a trace of nonabromobiphenyl (Neufeld et al., 1977).

Technical grade decabromobiphenyl produced in the US contains a minimum of 98% decabromobiphenyl and 2% nonabromobiphenyl (Neufeld et al., 1977).

2. Production, Use, Occurrence and Analysis

A review on the production, use and occurrence of PBBs has been published (Neufeld et al., 1977).

2.1 Production and use

(a) Production

The three commercial products, hexabromobiphenyl, octabromobiphenyl and decabromobiphenyl, were synthesized in an industrial research laboratory in the USA in 1969 (Head, 1974). PBBs are produced by bromination of biphenyl with bromine chloride in the presence of iron or a Friedel-Crafts catalyst in a closed vessel (Neufeld et al., 1977).

Hexabromobiphenyl

Commercial production of an undisclosed amount of hexabromobiphenyl (see preamble, p. 24) was first reported by one US company in 1970 (US Tariff Commission, 1972). The sole US producer ceased production in November 1974; they had produced an estimated 2.2 million kg in 1974 and about 5.1 million kg in the period 1970-1974. Production was halted after an incident in 1973 in which this chemical was accidentally mixed with cattle feed, leading to the quarantine and destruction of large numbers of cattle. The remaining stock of this product, held by the sole producer, was reportedly exhausted in April 1975 (Neufeld et al., 1977). There are no imports of hexabromobiphenyl to the US, except perhaps in finished products (Neufeld et al., 1977).

Hexabromobiphenyl is produced by one company in the Federal Republic of Germany (Neufeld et al., 1977).

Octabromobiphenyl and decabromobiphenyl

One US company is reported to have produced both chemicals in the period 1970-1976, and another produced decabromobiphenyl in the period 1973-1976. The estimated total quantity of the two chemicals produced by the two companies in the period 1970-1976 is 696 thousand kg; 365 thousand kg were produced in 1976 alone, all of which was exported. There are no US imports of these two products, except perhaps in finished products (Neufeld et al., 1977).

One company in the UK is reported to have produced decabromobiphenyl in the past but has discontinued production. One company in the Federal Republic of Germany produces octabromobiphenyl (Neufeld et al., 1977).

PBBs have never been produced in Japan; some was imported in the past, but none in recent years.

(b) Use

The principal PBB product to reach large-scale commercial production in the US was hexabromobiphenyl. Before its production was stopped in November 1974, this chemical was used as a fire retardant in three main commercial products: (1) acrylonitrile-butadiene-styrene (ABS) plastics (see IARC, 1978a); (2) coatings and lacquers; and (3) polyurethane foams. Octabromobiphenyl and decabromobiphenyl are also believed to be used as flame retardants, but no specific information was found on the kinds of products into which they were incorporated. No commercial use of these two products was reported in the US during 1970-1974 (Neufeld et al., 1977).

Of the estimated 2.2 million kg hexabromobiphenyl produced in 1974 (the final year of US production), about 900 thousand kg were used as a flame retardant in ABS plastic products. About 48% of the ABS resins, containing about 10% hexabromobiphenyl, was used in the fabrication of business machines and industrial equipment, such as typewriters, calculators, microfilm readers and business machine housings. Another 35% was used in electrical manufacturing, in the fabrication of radio and television parts, thermostats, electric razors and hand tool housings. About 12% was used in projector housings and cinema equipment cases. Approximately 1% was used in various small automotive parts, such as electrical wire connectors, switch connectors and speaker grilles; and about 4% was used in miscellaneous items, such as electrical appliances, motor housings and components for industrial equipment. The use of hexabromobiphenyl as a flame retardant in thermoplastics was confined to those products which do not come into contact with food or feed and which were not used in fabrics to which humans are exposed (Neufeld et al., 1977).

In 1974, about 34 thousand kg hexabromobiphenyl were used as flame retardant in cable coatings. It was also used as flame retardant in polyurethane foam for the automobile industry by at least two companies, in one of these until 1972; at that time, however, the automobile companies stopped accepting hexabromobiphenyl as a flame retardant chemical because it did not decompose in the ultimate incineration of scrapped automobiles (Neufeld et al., 1977).

All use of hexabromobiphenyl in the US was discontinued in 1974 because of the hazard to human health discovered after its accidental use in Michigan in 1973 (Neufeld et al., 1977).

Potential uses of PBBs include: in the synthesis of biphenyl esters; in a modified Würtz synthesis; as colour activators in light-sensitive compositions; as molecular weight control agents in polybutadiene; as wood preservatives; and as voltage stabilizing agents in electrical insulation (Neufeld et al., 1977).

In May 1974, the US Food and Drug Administration established the following enforcement limits for unavoidable residues of PBBs in foods: 1 mg/kg in the fat of meat, milk and dairy products; 0.3 mg/kg in animal feeds; and 0.1 mg/kg in eggs. These enforcement guidelines were reduced in November 1974 to 0.3 mg/kg in the fat of meat, milk and dairy products; and 0.05 mg/kg in eggs and animal feeds. In February 1977, the Food and Drug Administration rejected a petition to lower the enforcement guideline level to 0.02 mg/kg for all food products (Kolbye, 1977).

2.2 Occurrence

PBBs are not known to occur as natural products. Two reviews describe their pollution potential (Mumma & Wallace, 1975; Neufeld et al., 1977).

(a) Water and sediments

Concentrations of PBBs in a river near the plant where they were produced were found to decrease from 3.2 µg/l near the effluent discharge to 0.01 µg/l eight miles downstream (US Environmental Protection Agency, 1976). Sediments at the bottom of a canal into which discharges from one PBB manufacturer flow contained more than 250 mg/kg PBBs (Anon., 1977a).

(b) Soil and plants

The persistence of PBBs in soils and their uptake by plants have been studied experimentally in Brookston loam and Spinks sandy loam soils in Michigan, in manured soil, and in orchard grass and carrots. Sandy loams were analysed after being treated with 100 µg PBBs/ml acetone for 0, 6, 12 and 24 weeks: PBBs were found to persist for half a year in all samples. At a farm with a contaminated dairy herd (0.2-0.3 mg/l in the milk), the concentration of PBBs was measured in manured soil samples taken at three different locations in a field approximately one year after contamination. The major PBB component isomer (2,2',4,4',5,5'-hexabromobiphenyl) was found in concentrations of 4.1, 5.7 and 4.4 µg/kg. Orchard grass and carrots were grown for 5 and 11 weeks, respectively, in soil treated with powdered PBBs at concentrations of 0, 0.1, 1.0, 10 and 100 mg/kg: PBBs were not detected in any of the orchard grass or carrot tops but were detected in trace amounts in carrot roots grown in soil treated with 10 or 100 mg/kg (Jacobs et al., 1976).

(c) Food

In 1974, measurements made at 22 contaminated farm premises in Michigan during 1973 showed levels of PBBs (largely hexabromobiphenyl) in milk of as high as 595 mg/l and in eggs of 59.7 mg/kg. In a survey in March 1975 by the US Food and Drug Administration of 16 herds of contaminated Michigan dairy cattle, milk fat levels of PBBs were found to range from 1-13 mg/kg (Kay, 1977).

Milk from 4 dairy herds containing from 0.093 to 0.307 mg/kg PBBs (fat basis) was processed into a variety of finished dairy products: PBBs were concentrated in the high-fat products. Spray drying appeared to reduce the content of PBBs in whole milk and skim milk by 30-36% and 61-69%, respectively (Murata et al., 1977).

(d) Animals

In July 1973, cattle feed was contaminated with about 1000 kg of a hexabromobiphenyl[1] (Kimbrough et al., 1977; Meester & McCoy, 1976), resulting in contamination levels of from over 4000-13 500 mg/kg. The contaminated feed was used by farmers in Michigan between July 1973 and May 1974, until PBBs were identified as the toxic factor responsible for the ill effects, which were first reported in a dairy herd in September 1973. Body fat levels of 120 to 400 mg/kg were found in aborted calves. In October 1974, measurements made by the Michigan Department of Agriculture at 22 farm premises showed PBBs in poultry tissue and cattle tissue at levels of 4600 and up to 2700 mg/kg, respectively (Kay, 1977).

PBBs were found in 1974, 1976 and 1977 in wild ducks within two miles of a plant producing PBBs, at levels of 0.25, 0.29 and 1.8 mg/kg, respectively (Hesse & Powers, 1978).

(e) Marine organisms

Trace amounts of PBBs were found in three catfish taken from a river near a plastics plant where PBBs had formerly been used as flame retardants (Anon., 1977b). A carp fished near a plant where PBBs were produced contained 1.33 mg/kg (Hesse & Powers, 1978).

(f) Human exposure

Approximately 8000 (Kay, 1977) or 10 000-12 500 (Subcommittee on Polybrominated Biphenyls, 1977) Michigan residents, primarily farm families and their neighbours, were exposed in 1973 and 1974 to meat, milk and eggs contaminated with PBBs. At least 2000 families received the heaviest exposure (Meester & McCoy, 1976).

In a survey carried out by the Michigan Department of Public Health in 1974 on 165 persons from the area, blood levels of 0-2.26 mg/l PBBs were found. The PBB levels in the fat of breast milk from 95 Michigan

---

[1] The compound is believed to have been Firemaster FF-1.

nursing women in August 1976 ranged from 0-1.22 mg/kg (Subcommittee on Polybrominated Biphenyls, 1977). Levels of 0.058-273 mg/kg PBBs were found in the fat of a further 116 subjects examined (Meester & McCoy, 1976).

An unknown number of people were exposed occupationally, including dairymen exposed to contaminated feed dust, elevator and mill personnel exposed to PBB residues, and operators at the company where grinding and blending with the anticaking agent for flame retardant use was done.

Analysis of the blood of 7 employees of the company that manufactured PBBs showed levels of from 0.015 mg/l (three-months' exposure) to 0.085 mg/l (26-months' exposure) in 6 workers, and 0.006 mg/l in a supervisor who had worked 38 months (Kay, 1977).

2.3 Analysis

Some typical methods of analysis which have been employed to determine PBBs in environmental samples are summarized in Table 1.

The analysis of PBBs by high-performance liquid chromatography, gas chromatography/mass spectrometry, ultra-violet spectrometry and nuclear magnetic resonance spectrometry has been reviewed (De Kok et al., 1977). The AOAC (Association of Official Analytical Chemists) extraction technique, followed by gel-permeation chromatography and gas chromatography/electron-capture detector has been used to determine PBBs in milk and dairy products (Fehringer, 1975). Use of ultra-violet irradiation of extract followed by gas chromatography or mass spectrometry has been proposed to confirm the presence of PBBs (Erney, 1975; Murata et al., 1977; Trotter, 1977).

3. Biological Data Relevant to the Evaluation of Carcinogenic Risk to Humans

3.1 Carcinogenicity and related studies in animals[1]

Oral administration

Rat: Twenty male and 20 female 2½-month-old Sherman rats were given single doses of 1000 mg/kg bw PBBs (Firemaster FF-1) by gavage as a 5% solution in peanut oil. The rats were fasted for 15 hours prior to dosing; they were killed in groups of 5 at 2, 6, 10 and 14 months after treatment, and their organs were examined microscopically. Peanut oil was given to a

---

[1]The Working Group was aware of studies in progress to assess the carcinogenicity of Firemaster FF-1 by oral administration to rats (Kimbrough et al., 1978) and to mice (Subcommittee on Polybrominated Biphenyls, 1977).

Table 1

Sample type	ANALYTICAL METHOD			Reference
	Extraction clean-up	Detection	Limit of Detection	
Food				
Milk	Weigh sample, add ethanol, extract (petroleum ether-diethylether), centrifuge, recover ether phase, evaporate to dryness, column chromatography	GC/ECD[1]	1.4 µg/kg (ppb)	Willett et al. (1978)
Milk	Sample+water+potassium oxalate+ ethanol, extract (ethyl ether and petroleum ether), dry, evaporate to dryness, column chromatography	GC/ECD		Willett & Irving (1976)
Plant extracts	Extract (benzene, 2-propanol), wash with water to remove 2-propanol, dry, column chromatography	GC/ECD GC/MS[2] confirmation	10 µg/kg (ppb)	Jacobs et al. (1976)
Biological				
Bile and faeces; plasma	Weigh sample, add ethanol (except for faeces), extract (petroleum ether-diethylether), centrifuge, recover ether phase, evaporate to dryness, column chromatography	GC/ECD	1.4 µg/kg (ppb) 1 µg/kg (plasma)	Willett et al. (1978)
Faeces; animal tissues	Grind with sodium sulphate, extract (anhydrous ether and petroleum ether), filter, column chromatography	GC/ECD		Willett & Irving (1976)
Plasma	Homogenize with florisil, column chromatography			
Miscellaneous				
Soil	Extract (benzene, 2-propanol), wash with water to remove 2-propanol, dry, column chromatography	GC/ECD GC/MS confirmation	0.1 µg/kg (ppb)	Jacobs et al. (1976)

[1] GC/ECD = gas chromatography/electron capture detector
[2] GC/MS = gas chromatography/mass spectrometry

control group of 20 males and 20 females, which were examined in the same way as the test animals. None of the experimental rats but 2 control animals died before they were to be killed. No difference was noted in weight gain of controls and test animals. Neoplastic liver nodules, described as 'hyperplastic' were observed in 4/5 females killed 10 months after treatment and in 2/5 males and 3/5 females killed 14 months after treatment with PBBs. No such lesions were found among the controls. In addition, in the animals killed 6 or more months after treatment with PBBs, the hepatocytes in the centre of the liver lobules were enlarged, vacuolated or had foamy cytoplasm and a general pleomorphism and multinucleation (Kimbrough et al., 1977, 1978) [The Working Group noted the preliminary character of this study, the small number of animals and the short duration of the experiment].

3.2 Other relevant biological data

A report is available about studies in progress on the toxicity of PBBs (Subcommittee on Polybrominated Biphenyls, 1977).

(a) Experimental systems

Toxic effects

The acute oral LD_{50} of Firemaster BP-6 in rats was 21.5 g/kg bw (Kay, 1977). Subchronic dosing with Firemaster FF-1 reduced the lethal dose: doses of 100 mg/kg bw on 5 days/ week for a total of 22 doses killed 100% of female rats at 41-53 days and 38% of male rats at 50-73 days (Gupta & Moore, 1978). Guinea-pigs and mink exposed to commercial hexabromobiphenyl appeared to be more susceptible (Sleight & Sanger, 1976).

Toxic effects in rats and mice exposed to Firemaster FF-1 included bodyweight decrease, liver hypertrophy and fatty infiltration, hepatocyte swelling, cytoplasmic vacuoles and microscopic abscesses (Moore et al., 1978; Sleight & Sanger, 1976). Chloracne-like lesions occurred in rabbit ears exposed to a total dose of 60 mg Firemaster FF-1/ear (Kimbrough et al., 1977).

Suppression of cell-mediated immunity occurred in rats and mice given oral doses of Firemaster FF-1, as indicated by depressed responsiveness of splenic lymphocytes to mitogenic stimulation by polyclonal T-cell activators. Indications of humoral immune suppression were observed in mice (Luster et al., 1978).

Mice and rats exhibited neuromuscular dysfunction in a variety of tests after administration of Firemaster FF-1 or 2,2',4,4',5,5'-hexabromobiphenyl. Rats appeared to be more severely affected than mice; the effects in rats tended to worsen over 30 days after treatment (Tilson et al., 1978).

Firemaster BP-6 induced both cytochrome P-450 and P-448 in the liver of rats 192 hours after treatment with single i.p. injections of 25 mg/kg bw

(Dent, 1978; Dent *et al.*, 1976); the major component, 2,2',4,4',5,5'-hexabromobiphenyl, induced only cytochrome P-450 (Goldstein & Hickman, 1978).

Mice fed a diet containing Firemaster BP-6 were more susceptible to chlorinated hydrocarbon solvent-induced renal and hepatic damage and to the lethal effects of chloroform and carbon tetrachloride than mice that received a control diet (Kluwe *et al.*, 1978).

Firemaster FF-1 has also been reported to stimulate biliary secretion in rats and mice (Cagen & Gibson, 1977) and to induce porphyria in rats (Goldstein & Hickman, 1978).

The effects of Firemaster FF-1 in cattle included anorexia, decreased milk production, lameness, haematomas, abscesses, abnormal hoof growth, alopecia and skin thickening. Autopsy revealed liver and kidney degeneration (Jackson & Halbert, 1974). Glandular hyperplasia of the major intrahepatic bile ducts of the liver was seen in cows fed Firemaster BP-6 and of the gall bladder in sheep (Gutenmann & Lisk, 1975).

The acute oral LD_{50} in rats of a technical octabromobiphenyl (OBB), containing hepta-, nona- and decabromobiphenyls, was greater than 17 g/kg bw. In a 28-day study in which the rats were fed 100 or 1000 mg/kg of diet, the following effects were observed: marked hepatocyte enlargement, cytoplasmic inclusions, foamy cytoplasm and decreased hepatic glycogen. A dose-related accumulation of OBB in body fat was seen (Waritz *et al.*, 1977). Furthermore, thyroid hyperplasia was observed in rats fed OBB at levels of 1, 0.1 or 0.01% in the diet for 2 years; kidney lesions consisting of hyaline degenerative cytoplasmic changes were also seen with all levels (Norris *et al.*, 1975).

Embryotoxicity and teratogenicity

Teratogenic and embryotoxic effects of Firemaster BP-6 have been demonstrated in rats. A single dose of 40, 200, 400 or 800 mg/kg bw was administered by gavage on days 6-14 of pregnancy: doses of 400 or 800 mg/kg bw induced cleft palate and diaphragmatic hernia; the number of resorbed and dead foetuses increased markedly with administration of 800 mg on days 7-12; and maternal toxicity occurred in animals given 800 mg/kg (Beaudoin, 1977).

Firemaster BP-6 had a weak teratogenic activity in mice, producing exencephaly when fed to pregnant dams at 100 or 1000 mg/kg of diet on days 7-18 of pregnancy and cleft palates and hydronephrosis when given at 1000 mg/kg of diet (Corbett *et al.*, 1975). No embryotoxic or teratogenic effects were observed in the offspring of Sprague-Dawley rats given 0.25, 0.5, 1, 5 or 10 mg Firemaster BP-6 by gavage from days 7-15 of pregnancy (Harris *et al.*, 1978).

Abortion or foetal death were reported in the offspring of Holstein heifers fed Firemaster BP-6 in daily doses of 25 g/day (Durst et al., 1977). After the Michigan accident, many calves were born with extraordinarily large heads (Meester & McCoy, 1976).

Absorption, distribution, excretion and metabolism

^{14}C-Labelled 2,2',4,4',5,5'-hexabromobiphenyl, the major component of Firemaster BP-6, administered to male rats was readily absorbed from the intestine and distributed throughout the body; it was stored principally in fat. Extrapolation of the rate of excretion indicates that less than 10% of any dose would be excreted (Matthews et al., 1977). In Sprague-Dawley rats, concentrations of PBBs were higher in the livers of the offspring than in the livers of their mothers treated with 50 mg/kg of diet Firemaster BP-6 on days 8-21 of gestation. Transfer of PBBs via the milk was more important than placental transfer in relation to the occurrence of PBBs in newborns (Rickert et al., 1978).

PBB is excreted in the milk of ruminants exposed to Firemaster BP-6 (Gutenmann & Lisk, 1975). In a study on the placental transfer of Firemaster BP-6, detectable levels were found in tissues of 17 Holstein calves from dams that had been contaminated at least 7-9 months prior to delivery (Detering et al., 1975).

Mutagenicity and other short-term tests

Firemaster found in the grain mill where the Michigan contamination occurred was given orally to mice at doses of 50 and 500 mg/kg bw; no chromosome aberrations were induced in bone marrow cells (Wertz & Ficsor, 1978).

(b) Humans

No immediate effects on the health of people exposed to PBBs have been reported. They cross the placental barrier and are secreted in mothers' milk (Isbister, 1977). For a description of the occurrence of PBBs in human blood, milk and adipose tissue, see section 2.2, p. 112.

In a study of 933 exposed farmers and other residents of Michigan and of 229 Wisconsin farmers who were not exposed, the Michigan population had a significantly higher prevalence of skin, neurological or musculoskeletal symptoms (Anderson et al., 1978a). The authors concluded that 'the existing differences observed could not be explained without considering an etiological role for exposure to PBB'. An increased incidence of abnormal serum glutamic-oxaloacetic transaminase and serum glutamic-pyruvic transaminase was also found among Michigan dairy farm residents and consumers of products from these farms, compared with Wisconsin dairy farm residents (Anderson et al., 1978b). In a study of 55 employees of the Michigan Chemical Corporation, where Firemaster BP-6 was manufactured from 1970-1974, there was an increased prevalence of chest and skin symptoms compared with farmers exposed to PBBs (Anderson et al., 1978c).

Abnormal immune function of peripheral lymphocytes was also observed in some Michigan residents. These alterations included an increased number of null lymphocytes, decreased response to mitogens and decreased response of mixed lymphocyte cultures (Bekesi et al., 1978). Barr (1978) stated that 'many Michigan farm children are considered as having a deterioration in health coincident with contamination of dairy cattle by PBBs; a discrete syndrome was not identified'.

3.3 Case reports and epidemiological studies[1]

No data were available to the Working Group.

4. Summary of Data Reported and Evaluation

4.1 Experimental data

Only one commercial preparation of polybrominated biphenyls (Firemaster FF-1), composed primarily of penta-, hexa- and heptabrominated congeners, was tested in one experiment in rats by oral administration of single doses. Although neoplastic changes of the liver were reported, the study was considered to be inadequate.

Firemaster FF-1 is embryotoxic and teratogenic. Cytogenetic tests in mice were negative.

4.2 Human data

No case reports or epidemiological studies were available to the Working Group.

The extensive production of polybrominated biphenyls during the last decade, their persistence in the environment and in the body, and their use, mainly as flame retardants, indicate that widespread human exposure occurs. This is confirmed by their presence in water and food and in human tissues and body fluids.

4.3 Evaluation

No evaluation of the carcinogenicity of polybrominated biphenyls could be made on the basis of the available data.

---

[1]The Working Group was aware of an epidemiological study in progress in Michigan residents (IARC, 1978b)

5. References

Anderson, H.A., Lilis, R., Selikoff, I.J., Rosenman, K.D., Valciukas, J.A. & Freedman, S. (1978a) Unanticipated prevalence of symptoms among dairy farmers in Michigan and Wisconsin. Environ. Health Perspect., 23, 217-226

Anderson, H.A., Holstein, E.C., Daum, S.M., Sarkozi, L. & Selikoff, I.J. (1978b) Liver function tests among Michigan and Wisconsin dairy farmers. Environ. Health Perspect., 23, 333-339

Anderson, H.A., Wolff, M.S., Fischbein, A. & Selikoff, I.J. (1978c) Investigation of the health status of Michigan Chemical Corporation employees. Environ. Health Perspect., 23, 187-191

Anon. (1977a) Second NJ firm making PBBs doesn't have a water discharge permit. Pestic. Toxic Chem. News, 5, 35

Anon. (1977b) Ohio River catfish had only traces of PBBs, EPA says, contrary to preliminary reports. Air Water Pollut. Rep., September 19, p. 376

Barr, M., Jr (1978) Pediatric health aspects of PBBs. Environ. Health Perspect., 23, 291-294

Beaudoin, A.R. (1977) Teratogenicity of polybrominated biphenyls in rats. Environ. Res., 14, 81-86

Bekesi, J.G., Holland, J.F., Anderson, H.A., Fischbein, A.S., Rom, W., Wolff, M.S. & Selikoff, I.J. (1978) Lymphocyte function of Michigan dairy farmers exposed to polybrominated biphenyls. Science, 199, 1207-1209

Cagen, S.Z. & Gibson, J.E. (1977) Stimulation of biliary function following polybrominated biphenyls (Abstract No. 39). Toxicol. appl. Pharmacol., 41, 147

Corbett, T.H., Beaudoin, A.R., Cornell, R.G., Anver, M.R., Schumacher, R., Endres, J. & Szwabowska, M. (1975) Toxicity of polybrominated biphenyls (Firemaster BP-6) in rodents. Environ. Res., 10, 390-396

De Kok, J.J., De Kok, A., Brinkman, U.A.T. & Kok, R.M. (1977) Analysis of polybrominated biphenyls. J. Chromatogr., 142, 367-383

Dent, J.G. (1978) Characteristics of cytochrome P-450 and mixed function oxidase enzymes following treatment with PBBs. Environ. Health Perspect., 23, 301-307

Dent, J.G., Netter, K.J. & Gibson, J.E. (1976) The induction of hepatic microsomal metabolism in rats following acute administration of a mixture of polybrominated biphenyls. Toxicol. appl. Pharmacol., 38, 237-249

Detering, C.N., Prewitt, L.R., Cook, R.M. & Fries, G.F. (1975) Placental transfer of polybrominated biphenyl by Holstein cows (Abstract No. P94). J. Dairy Sci., 58, 764-765

Durst, H.I., Willett, L.B., Brumm, C.J. & Mercer, H.D. (1977) Effects of polybrominated biphenyls on health and performance of pregnant Holstein heifers. J. Dairy Sci., 60, 1294-1300

Erney, D.R. (1975) Confirmation of polybrominated biphenyl residues in feeds and dairy products, using an ultraviolet irradiation-gas-liquid chromatographic technique. J. Assoc. off. anal. Chem., 58, 1202-1205

Fehringer, N.V. (1975) Determination of polybrominated biphenyl residues in dairy products. J. Assoc. off. anal. Chem., 58, 978-982

Goldstein, J.A. & Hickman, P. (1978) Comparison of a commercial polybrominated mixture (Firemaster BP-6) with 2,4,5,2',4',5'-hexachlorobiphenyl and a tetrabromonaphthalene as inducers of hepatic mixed-function oxidases. Toxicol. appl. Pharmacol. (in press)

Gupta, B.N. & Moore, J.A. (1978) Toxicity of Firemaster FF-1 in the rat. In: Abstracts of the 115th Annual American Veterinary Medical Association Meeting, Dallas, Texas, 1978

Gutenmann, W.H. & Lisk, D.J. (1975) Tissue storage and excretion in milk of polybrominated biphenyls in ruminants. J. agric. Food Chem., 23, 1005-1007

Harris, S.J., Cecil, H.C. & Bitman, J. (1978) Embryotoxic effects of polybrominated biphenyls (PBB) in rats. Environ. Health Perspect., 23, 295-300

Hass, J.R., McConnell, E.E. & Harvan, D.J. (1978) Chemical and toxicologic evaluation of Firemaster BP-6. Agric. Food Chem., 26, 94-99

Head, J.D. (1974) A Case Study. Polybrominated Biphenyls, Contract No. 68-01-2262, US Environmental Protection Agency, Washington DC

Hesse, J.L. & Powers, R.A. (1978) Polybrominated biphenyl (PBB) contamination of the Pine River, Gratiot and Midland Counties, Michigan. Environ. Health Perspect., 23, 19-25

IARC (1978a) IARC Monographs on the Evaluation of the Carcinogenic Risk of Chemicals to Humans, 19, Plastics and Synthetic Elastomers - Some Monomers, Homopolymers and Copolymers (in press)

IARC (1978b) <u>Directory of On-Going Research in Cancer Epidemiology, 1978</u>, Lyon (<u>IARC Scientific Publications No. 26</u>), pp. 249-250 (Abstract No. 655)

Isbister, J.L. (1977) PBBs in human health. <u>Clin. Med.</u>, <u>84</u>, 22-24

Jackson, T.F. & Halbert, F.L. (1974) A toxic syndrome associated with the feeding of polybrominated biphenyl-contaminated protein concentrate to dairy cattle. <u>J. Am. vet. med. Assoc.</u>, <u>165</u>, 437-439

Jacobs, L.W., Chow, S.-F. & Tiedje, J.M. (1976) Fate of polybrominated biphenyls (PBB's) in soils. Persistence and plant uptake. <u>J. agric. Food Chem.</u>, <u>24</u>, 1198-1201

Kay, K. (1977) Polybrominated biphenyls (PBB) environmental contamination in Michigan, 1973-1976. <u>Environ. Res.</u>, <u>13</u>, 74-93

Kimbrough, R.D., Burse, V.W., Liddle, J.A. & Fries, G.F. (1977) Toxicity of polybrominated biphenyl. <u>Lancet</u>, <u>ii</u>, 602-603

Kimbrough, R.D., Burse, V.W. & Liddle, J.A. (1978) Persistent liver lesions in rats after a single oral dose of polybrominated biphenyls (FireMaster FF-1) and concomitant PBB tissue levels. <u>Environ. Health Perspect.</u>, <u>23</u>, 265-273

Kluwe, W.M., McCormack, K.M. & Hook, J.B. (1978) Potentiation of hepatic and renal toxicity of various compounds by prior exposure to polybrominated biphenyls. <u>Environ. Health Perspect.</u>, <u>23</u>, 241-246

Kolbye, A.C., Jr (1977) <u>Statement on Polybrominated Biphenyls (PBB's) before the Subcommittee on Space, Technology and Science, Committee on Commerce, Space and Transportation, US Senate, March 31</u>

Luster, M.I., Faith, R.E. & Moore, J.A. (1978) Effects of polybrominated biphenyls (PBB) on immune response in rodents. <u>Environ. Health Perspect.</u>, <u>23</u>, 227-232

Matthews, H.B., Kato, S., Morales, N.M. & Tuey, D.B. (1977) Distribution and excretion of 2,4,5,2',4',5'-hexabromobiphenyl, the major component of Firemaster BP-6. <u>J. Toxicol. environ. Health</u>, <u>3</u>, 599-605

Meester, W.D. & McCoy, D.J., Sr (1976) <u>Human toxicology of polybrominated biphenyls</u>. In: <u>Management of the Poisoned Patient</u>, Princeton, New Jersey, Science Press, pp. 32-61

Moore, J.A., Luster, M.I., Gupta, B.N. & McConnell, E.E. (1978) Toxicological and immunological effects of a commercial polybrominated biphenyl mixture (Firemaster FF-1). <u>Toxicol. appl. Pharmacol.</u> (in press)

Mumma, C.E. & Wallace, D.D. (1975) Survey of Industrial Processing Data, Task II, Pollution Potential of Polybrominated Biphenyls, EPA-560/3-75-004, Springfield, Virginia, National Technical Information Service

Murata, T., Zabik, M.E. & Zabik, M. (1977) Polybrominated biphenyls in raw milk and processed dairy products. J. Dairy Sci., 60, 516-520

Neufeld, M.L., Sittenfield, M. & Wolk, K.F. (1977) Market Input/Output Studies, Task IV, Polybrominated Biphenyls, EPA-560/6-77-017, Springfield, Virginia, National Technical Information Service

Norris, J.M., Ehrmantraut, J.W., Gibbons, C.L., Kociba, R.J., Schwetz, B.A., Rose, J.Q., Humiston, C.G., Jewett, G.L., Crummett, W.B., Gehring, P.J., Tirsell, J.B. & Brosier, J.S. (1973) Toxicological and environmental factors involved in the selection of decabromodiphenyl oxide as a fire retardant chemical. In: Applied Polymer Symposium No. 22, New York, John Wiley & Sons, pp. 195-219

Norris, J.M., Kociba, R.J., Schwetz, B.A., Rose, J.Q., Humiston, C.G., Jewett, G.L., Gehring, P.J. & Mailhes, J.B. (1975) Toxicology of octabromobiphenyl and decabromodiphenyl oxide. Environ. Health Perspect., 11, 153-161

O'Keefe, P.W. (1978) Formation of brominated dibenzofurans from pyrolysis of the polybrominated biphenyl fire retardant, FireMaster FF-1. Environ. Health Perspect., 23, 347-350

Rickert, D.E., Dent, J.G., Cagen, S.Z., McCormack, K.M., Melrose, P. & Gibson, J.E. (1978) Distribution of polybrominated biphenyls after dietary exposure in pregnant and lactating rats and their offspring. Environ. Health Perspect., 23, 63-66

Sleight, S.D. & Sanger, V.L. (1976) Pathologic features of polybrominated biphenyl toxicosis in the rat and guinea pig. J. Am. vet. med. Assoc., 169, 1231-1235

Subcommittee on the Health Effects of Polychlorinated Biphenyls and Polybrominated Biphenyls (1976) Final Report, July 1976, Washington DC, Department of Health, Education and Welfare

Subcommittee on the Health Effects of Polychlorinated Biphenyls and Polybrominated Biphenyls (1978) Final Report of the Subcommittee of the Health Effects of Polychlorinated Biphenyls and Polybrominated Biphenyls of the DHEW Committee to coordinate toxicology and related programs. Environ. Health Perspect., 24, 129-239

Subcommittee on Polybrominated Biphenyls (1977) DHEW Committee to Coordinate Toxicology and Related Programs, April 28, 1977, Research Triangle Park, North Carolina, National Institute of Environmental Health Sciences

Tilson, H.A., Cabe, P.A. & Mitchell, C.L. (1978) Behavioral and neurological toxicity of polybrominated biphenyls in rats and mice. Environ. Health Perspect., 23, 257-263

Trotter, W.J. (1977) Confirming low levels of hexabromobiphenyl by gas-liquid chromatography of photolysis products. Bull. environ. Contam. Toxicol., 18, 726-733

US Environmental Protection Agency (1976) Polybrominated biphenyls (PBB's). Summary Characterizations of Selected Chemicals of Near-Term Interest, EPA 560/4-76-004, Springfield, Virginia, National Technical Information Service, pp. 41-43

US Tariff Commission (1972) Synthetic Organic Chemicals, US Production and Sales, 1970, TC Publication 479, Washington DC, US Government Printing Office, p. 41

Waritz, R.S., Aftosmis, J.G., Culik, R., Dashiell, O.L., Faunce, M.M., Griffith, F.D., Hornberger, C.S., Lee, K.P., Sherman, H. & Tayfun, F.O. (1977) Toxicological evaluations of some brominated biphenyls. Am. ind. Hyg. Assoc. J., 38, 307-320

Wertz, G.F. & Ficsor, G. (1978) Cytogenetic and teratogenic test of polybrominated biphenyls in rodents. Environ. Health Perspect., 23, 129-132

Willett, L.B. & Irving, H.A. (1976) Distribution and clearance of polybrominated biphenyls in cows and calves. J. Dairy Sci., 59, 1429-1439

Willett, L.B., Brumm, C.J. & Williams, C.L. (1978) Method for extraction, isolation, and detection of free polybrominated biphenyls (PBBs) from plasma, faeces, milk, and bile using disposable glassware. J. agric. Food Chem., 26, 122-125

SUPPLEMENTARY CORRIGENDA TO VOLUMES 1 - 17

Corrigenda covering Volumes 1 - 6 appeared in Volume 7, others appeared in Volumes 8, 10, 11, 12, 13, 15, 16 and 17.

Volume 4

p. 175   1st para. lines 3 and 8   *replace* 'maleic anhydride' *by* 'maleic hydrazide'

Volume 11

p. 135   para. 4 lines 9-11   *delete from* 'The maximum ...' *to end*

p. 138   ref. 11   *delete reference* 'Oser *et al.*'

IARC Monographs on the Evaluation of the Carcinogenic Risk of Chemicals to Humans (1978), Volume 18

CUMULATIVE INDEX TO IARC MONOGRAPHS ON THE EVALUATION
OF THE CARCINOGENIC RISK OF CHEMICALS TO HUMANS

Numbers underlined indicate volume, and numbers in italics indicate page. References to corrigenda are given in parentheses.

A

Acetamide	7,*197*
Acridine orange	16,*145*
Acriflavinium chloride	13,*31*
Actinomycins	10,*29*
Adriamycin	10,*43*
Aflatoxins	1,*145* (corr. 7,*319*)
	(corr. 8,*349*)
	10,*51*
Aldrin	5,*25*
Amaranth	8,*41*
5-Aminoacenaphthene	16,*243*
para-Aminoazobenzene	8,*53*
ortho-Aminoazotoluene	8,*61* (corr. 11,*295*)
para-Aminobenzoic acid	16,*249*
4-Aminobiphenyl	1,*74* (corr. 10,*343*)
2-Amino-5-(5-nitro-2-furyl)-1,3,4-thiadiazole	7,*143*
4-Amino-2-nitrophenol	16,*43*
Amitrole	7,*31*
Anaesthetics, volatile	11,*285*
Anthranilic acid	16,*265*
Aniline	4,*27* (corr. 7,*320*)
Apholate	9,*31*
Aramite®	5,*39*

Arsenic and inorganic arsenic compounds
  Arsenic pentoxide
  Arsenic trioxide
  Calcium arsenate
  Calcium arsenite
  Potassium arsenate
  Potassium arsenite
	2,*48*

Sodium arsenate
Sodium arsenite

Asbestos	2,17	(corr. 7,319)
	14	(corr. 15,341)
		(corr. 17,351)

    Actinolite
    Amosite
    Anthophyllite
    Chrysotile
    Crocidolite
    Tremolite

Auramine	1,69	(corr. 7,319)
Aurothioglucose	13,39	
Azaserine	10,73	
Aziridine	9,37	
2-(1-Aziridinyl)ethanol	9,47	
Aziridyl benzoquinone	9,51	
Azobenzene	8,75	

B

Benz[c]acridine	3,241	
Benz[a]anthracene	3,45	
Benzene	7,203	(corr. 11,295)
Benzidine	1,80	
Benzo[b]fluoranthene	3,69	
Benzo[j]fluoranthene	3,82	
Benzo[a]pyrene	3,91	
Benzo[e]pyrene	3,137	
Benzyl chloride	11,217	(corr. 13,243)
Benzyl violet 4B	16,153	
Beryllium and beryllium compounds	1,17	

    Beryl ore
    Beryllium oxide
    Beryllium phosphate
    Beryllium sulphate

BHC (technical grades)	5,47
Bis(1-aziridinyl)morpholinophosphine sulphide	9,55
Bis(2-chloroethyl)ether	9,117

N,N-Bis(2-chloroethyl)-2-naphthylamine	<u>4</u>,*119*
1,2-Bis(chloromethoxy)ethane	<u>15</u>,*31*
1,4-Bis(chloromethoxymethyl)benzene	<u>15</u>,*37*
Bis(chloromethyl)ether	<u>4</u>,*231* (corr. <u>13</u>,*243*)
Blue VRS	<u>16</u>,*163*
Brilliant blue FCF diammonium and disodium salts	<u>16</u>,*171*
1,4-Butanediol dimethanesulphonate (Myleran)	<u>4</u>,*247*
β-Butyrolactone	<u>11</u>,*225*
γ-Butyrolactone	<u>11</u>,*231*

C

Cadmium and cadmium compounds	<u>2</u>,*74* <u>11</u>,*39*
Cadmium acetate	
Cadmium carbonate	
Cadmium chloride	
Cadmium oxide	
Cadmium powder	
Cadmium sulphate	
Cadmium sulphide	
Cantharidin	<u>10</u>,*79*
Carbaryl	<u>12</u>,*37*
Carbon tetrachloride	<u>1</u>,*53*
Carmoisine	<u>8</u>,*83*
Catechol	<u>15</u>,*155*
Chlorambucil	<u>9</u>,*125*
Chloramphenicol	<u>10</u>,*85*
Chlorinated dibenzodioxins	<u>15</u>,*41*
Chlormadinone acetate	<u>6</u>,*149*
Chlorobenzilate	<u>5</u>,*75*
Chloroform	<u>1</u>,*61*
Chloromethyl methyl ether	<u>4</u>,*239*
Chloropropham	<u>12</u>,*55*
Chloroquine	<u>13</u>,*47*
para-Chloro-*ortho*-toluidine and its hydrochloride	<u>16</u>,*277*
Cholesterol	<u>10</u>,*99*

Chromium and inorganic chromium compounds   <u>2</u>,*100*
  Barium chromate
  Calcium chromate
  Chromic chromate
  Chromic oxide
  Chromium acetate
  Chromium carbonate
  Chromium dioxide
  Chromium phosphate
  Chromium trioxide
  Lead chromate
  Potassium chromate
  Potassium dichromate
  Sodium chromate
  Sodium dichromate
  Strontium chromate
  Zinc chromate hydroxide

Chrysene   <u>3</u>,*159*

Chrysoidine   <u>8</u>,*91*

C.I. Disperse Yellow 3   <u>8</u>,*97*

Cinnamyl anthranilate   <u>16</u>,*287*

Citrus Red No. 2   <u>8</u>,*101*

Copper 8-hydroxyquinoline   <u>15</u>,*103*

Coumarin   <u>10</u>,*113*

Cycasin   <u>1</u>,*157* (corr. <u>7</u>,*319*)
<u>10</u>,*121*

Cyclochlorotine   <u>10</u>,*139*

Cyclophosphamide   <u>9</u>,*135*

D

2,4-D and esters   <u>15</u>,*111*

D & C Red No. 9   <u>8</u>,*107*

Daunomycin   <u>10</u>,*145*

DDT and associated substances   <u>5</u>,*83* (corr. <u>7</u>,*320*)
  DDD (TDE)
  DDE

Diacetylaminoazotoluene   <u>8</u>,*113*

N,N'-Diacetylbenzidine   <u>16</u>,*293*

Diallate   <u>12</u>,*69*

2,4-Diaminoanisole and its sulphate   <u>16</u>,*51*

4,4'-Diaminodiphenyl ether	16,301
1,2-Diamino-4-nitrobenzene	16,63
1,4-Diamino-2-nitrobenzene	16,73
2,6-Diamino-3-(phenylazo)pyridine and its hydrochloride	8,117
2,4-Diaminotoluene	16,83
2,5-Diaminotoluene and its sulphate	16,97
Diazepam	13,57
Diazomethane	7,223
Dibenz[a,h]acridine	3,247
Dibenz[a,j]acridine	3,254
Dibenz[a,h]anthracene	3,178
7H-Dibenzo[c,g]carbazole	3,260
Dibenzo[h,rst]pentaphene	3,197
Dibenzo[a,e]pyrene	3,201
Dibenzo[a,h]pyrene	3,207
Dibenzo[a,i]pyrene	3,215
Dibenzo[a,l]pyrene	3,224
1,2-Dibromo-3-chloropropane	15,139
ortho-Dichlorobenzene	7,231
para-Dichlorobenzene	7,231
3,3'-Dichlorobenzidine	4,49
trans-1,4-Dichlorobutene	15,149
3,3'-Dichloro-4,4'-diaminodiphenyl ether	16,309
Dieldrin	5,125
Diepoxybutane	11,115
1,2-Diethylhydrazine	4,153
Diethylstilboestrol	6,55
Diethyl sulphate	4,277
Diglycidyl resorcinol ether	11,125
Dihydrosafrole	1,170 10,233
Dihydroxybenzenes	15,155
Dimethisterone	6,167

Dimethoxane	15,177
3,3'-Dimethoxybenzidine (o-Dianisidine)	4,41
para-Dimethylaminoazobenzene	8,125
para-Dimethylaminobenzenediazo sodium sulphonate	8,147
trans-2[(Dimethylamino)methylimino]-5-[2-(5-nitro-2-furyl)vinyl]-1,3,4-oxadiazole	7,147
3,3'-Dimethylbenzidine (o-Tolidine)	1,87
Dimethylcarbamoyl chloride	12,77
1,1-Dimethylhydrazine	4,137
1,2-Dimethylhydrazine	4,145 (corr. 7,320)
Dimethyl sulphate	4,271
Dinitrosopentamethylenetetramine	11,241
1,4-Dioxane	11,247
2,4'-Diphenyldiamine	16,313
Disulfiram	12,85
Dithranol	13,75
Dulcin	12,97

E

Endrin	5,157
Eosin and its disodium salt	15,183
Epichlorohydrin	11,131 (corr. 18,125)
1-Epoxyethyl-3,4-epoxycyclohexane	11,141
3,4-Epoxy-6-methylcyclohexylmethyl-3,4-epoxy-6-methylcyclohexane carboxylate	11,147
cis-9,10-Epoxystearic acid	11,153
Ethinyloestradiol	6,77
Ethionamide	13,83
Ethylene dibromide	15,195
Ethylene oxide	11,157
Ethylene sulphide	11,257
Ethylenethiourea	7,45
Ethyl methanesulphonate	7,245
Ethyl selenac	12,107

Ethyl tellurac	<u>12</u>,*115*
Ethynodiol diacetate	<u>6</u>,*173*
Evans blue	<u>8</u>,*151*

F

Fast green FCF	<u>16</u>,*187*
Ferbam	<u>12</u>,*121* (corr. <u>13</u>,*243*)
2-(2-Formylhydrazino)-4-(5-nitro-2-furyl)thiazole	<u>7</u>,*151* (corr. <u>11</u>,*295*)
Fusarenon-X	<u>11</u>,*169*

G

Glycidaldehyde	<u>11</u>,*175*
Glycidyl oleate	<u>11</u>,*183*
Glycidyl stearate	<u>11</u>,*187*
Griseofulvin	<u>10</u>,*153*
Guinea green B	<u>16</u>,*199*

H

Haematite	<u>1</u>,*29*
Heptachlor and its epoxide	<u>5</u>,*173*
Hexamethylphosphoramide	<u>15</u>,*211*
Hycanthone and its mesylate	<u>13</u>,*91*
Hydrazine	<u>4</u>,*127*
Hydroquinone	<u>15</u>,*155*
4-Hydroxyazobenzene	<u>8</u>,*157*
8-Hydroxyquinoline	<u>13</u>,*101*
Hydroxysenkirkine	<u>10</u>,*265*

I

Indeno[1,2,3-*cd*]pyrene	<u>3</u>,*229*
Iron-dextran complex	<u>2</u>,*161*
Iron-dextrin complex	<u>2</u>,*161* (corr. <u>7</u>,*319*)
Iron oxide	<u>1</u>,*29*
Iron sorbitol-citric acid complex	<u>2</u>,*161*
Isatidine	<u>10</u>,*269*

Isonicotinic acid hydrazide	4,159
Isopropyl alcohol	15,223
Isopropyl oils	15,223
Isosafrole	1,169
	10,232

J

Jacobine	10,275

L

Lasiocarpine	10,281
Lead salts	1,40 (corr. 7,319)
	(corr. 8,349)
Lead acetate	
Lead arsenate	
Lead carbonate	
Lead phosphate	
Lead subacetate	
Ledate	12,131
Light green SF	16,209
Lindane	5,47
Luteoskyrin	10,163

M

Magenta	4,57 (corr. 7,320)
Maleic hydrazide	4,173 (corr. 18,125)
Maneb	12,137
Mannomustine and its dihydrochloride	9,157
Medphalan	9,167
Medroxyprogesterone acetate	6,157
Melphalan	9,167
Merphalan	9,167
Mestranol	6,87
Methoxychlor	5,193
2-Methylaziridine	9,61
Methylazoxymethanol acetate	1,164
	10,131

Methyl carbamate	<u>12</u>,*151*
N-Methyl-*N*,4-dinitrosoaniline	<u>1</u>,*141*
4,4'-Methylene bis(2-chloroaniline)	<u>4</u>,*65*
4,4'-Methylene bis(2-methylaniline)	<u>4</u>,*73*
4,4'-Methylenedianiline	<u>4</u>,*79* (corr. <u>7</u>,*320*)
Methyl iodide	<u>15</u>,*245*
Methyl methanesulphonate	<u>7</u>,*253*
N-Methyl-*N*'-nitro-*N*-nitrosoguanidine	<u>4</u>,*183*
Methyl red	<u>8</u>,*161*
Methyl selenac	<u>12</u>,*161*
Methylthiouracil	<u>7</u>,*53*
Metronidazole	<u>13</u>,*113*
Mirex	<u>5</u>,*203*
Mitomycin C	<u>10</u>,*171*
Monocrotaline	<u>10</u>,*291*
Monuron	<u>12</u>,*167*
5-(Morpholinomethyl)-3-[(5-nitrofurfurylidene)-amino]-2-oxazolidinone	<u>7</u>,*161*
Mustard gas	<u>9</u>,*181* (corr. <u>13</u>,*243*)

N

1-Naphthylamine	<u>4</u>,*87* (corr. <u>8</u>,*349*)
2-Naphthylamine	<u>4</u>,*97*
Native carrageenans	<u>10</u>,*181* (corr. <u>11</u>,*295*)
Nickel and nickel compounds	<u>2</u>,*126* (corr. <u>7</u>,*319*) <u>11</u>,*75*
Nickel acetate	
Nickel carbonate	
Nickel carbonyl	
Nickelocene	
Nickel oxide	
Nickel powder	
Nickel subsulphide	
Nickel sulphate	
Niridazole	<u>13</u>,*123*
5-Nitroacenaphthene	<u>16</u>,*319*
4-Nitrobiphenyl	<u>4</u>,*113*

5-Nitro-2-furaldehyde semicarbazone — 7,171

1[(5-Nitrofurfurylidene)amino]-2-imidazolidinone — 7,181

N-[4-(5-Nitro-2-furyl)-2-thiazolyl]acetamide — 1,181; 7,185

Nitrogen mustard and its hydrochloride — 9,193

Nitrogen mustard N-oxide and its hydrochloride — 9,209

N-Nitrosodi-n-butylamine — 4,197; 17,51

N-Nitrosodiethanolamine — 17,77

N-Nitrosodiethylamine — 1,107 (corr. 11,295); 17,83

N-Nitrosodimethylamine — 1,95; 17,125

N-Nitrosodi-n-propylamine — 17,177

N-Nitroso-n-ethylurea — 1,135; 17,191

N-Nitrosofolic acid — 17,217

N-Nitrosohydroxyproline — 17,303

N-Nitrosomethylethylamine — 17,221

N-Nitroso-N-methylurea — 1,125; 17,227

N-Nitroso-N-methylurethane — 4,211

N-Nitrosomethylvinylamine — 17,257

N-Nitrosomorpholine — 17,263

N'-Nitrosonornicotine — 17,281

N-Nitrosopiperidine — 17,287

N-Nitrosoproline — 17,303

N-Nitrosopyrrolidine — 17,313

N-Nitrososarcosine — 17,327

Norethisterone and its acetate — 6,179

Norethynodrel — 6,191

Norgestrel — 6,201

O

Ochratoxin A — 10,191

Oestradiol-17β — 6,99

Oestradiol mustard	<u>9</u>,*217*
Oestriol	<u>6</u>,*117*
Oestrone	<u>6</u>,*123*
Oil Orange SS	<u>8</u>,*165*
Orange I	<u>8</u>,*173*
Orange G	<u>8</u>,*181*
Oxazepam	<u>13</u>,*58*
Oxymetholone	<u>13</u>,*131*
Oxyphenbutazone	<u>13</u>,*185*

P

Parasorbic acid	<u>10</u>,*199*
Patulin	<u>10</u>,*205*
Penicillic acid	<u>10</u>,*211*
Phenacetin	<u>13</u>,*141*
Phenicarbazide	<u>12</u>,*177*
Phenobarbital	<u>13</u>,*157*
Phenobarbital sodium	<u>13</u>,*159*
Phenoxybenzamine and its hydrochloride	<u>9</u>,*223*
Phenylbutazone	<u>13</u>,*183*
meta-Phenylenediamine and its hydrochloride	<u>16</u>,*111*
para-Phenylenediamine and its hydrochloride	<u>16</u>,*125*
N-Phenyl-2-naphthylamine	<u>16</u>,*325*
Phenytoin	<u>13</u>,*201*
Phenytoin sodium	<u>13</u>,*202*
Polybrominated biphenyls	<u>18</u>,*107*
Polychlorinated biphenyls	<u>7</u>,*261* <u>18</u>,*43*
Ponceau MX	<u>8</u>,*189*
Ponceau 3R	<u>8</u>,*199*
Ponceau SX	<u>8</u>,*207*
Potassium bis(2-hydroxyethyl)dithiocarbamate	<u>12</u>,*183*
Progesterone	<u>6</u>,*135*
Pronetalol hydrochloride	<u>13</u>,*227* (corr. <u>16</u>,*387*)

1,3-Propane sultone	4,253 (corr. 13,243)
Propham	12,189
β-Propiolactone	4,259 (corr. 15,341)
n-Propyl carbamate	12,201
Propylene oxide	11,191
Propylthiouracil	7,67
Pyrimethamine	13,233

Q

para-Quinone	15,255
Quintozene (Pentachloronitrobenzene)	5,211

R

Reserpine	10,217
Resorcinol	15,155
Retrorsine	10,303
Rhodamine B	16,221
Rhodamine 6G	16,233
Riddelliine	10,313

S

Saccharated iron oxide	2,161
Safrole	1,169
	10,231
Scarlet red	8,217
Selenium and selenium compounds	9,245
Semicarbazide and its hydrochloride	12,209 (corr. 16,387)
Seneciphylline	10,319
Senkirkine	10,327
Sodium diethyldithiocarbamate	12,217
Soot, tars and shale oils	3,22
Sterigmatocystin	1,175
	10,245
Streptozotocin	4,221
	17,337

Styrene oxide	<u>11</u>,*201*
Succinic anhydride	<u>15</u>,*265*
Sudan I	<u>8</u>,*225*
Sudan II	<u>8</u>,*233*
Sudan III	<u>8</u>,*241*
Sudan brown RR	<u>8</u>,*249*
Sudan red 7B	<u>8</u>,*253*
Sunset yellow FCF	<u>8</u>,*257*

T

2,4,5-T and esters	<u>15</u>,*273*
Tannic acid	<u>10</u>,*253* (corr. <u>16</u>,*387*)
Tannins	<u>10</u>,*254*
Terpene polychlorinates (Strobane®)	<u>5</u>,*219*
Testosterone	<u>6</u>,*209*
Tetraethyllead	<u>2</u>,*150*
Tetramethyllead	<u>2</u>,*150*
Thioacetamide	<u>7</u>,*77*
4,4'-Thiodianiline	<u>16</u>,*343*
Thiouracil	<u>7</u>,*85*
Thiourea	<u>7</u>,*95*
Thiram	<u>12</u>,*225*
ortho-Toluidine and its hydrochloride	<u>16</u>,*349*
Trichloroethylene	<u>11</u>,*263*
Trichlorotriethylamine hydrochloride	<u>9</u>,*229*
Triethylene glycol diglycidyl ether	<u>11</u>,*209*
Tris(aziridinyl)-*para*-benzoquinone	<u>9</u>,*67*
Tris(1-aziridinyl)phosphine oxide	<u>9</u>,*75*
Tris(1-aziridinyl)phosphine sulphide	<u>9</u>,*85*
2,4,6-Tris(1-aziridinyl)-*s*-triazine	<u>9</u>,*95*
1,2,3-Tris(chloromethoxy)propane	<u>15</u>,*301*
Tris(2-methyl-1-aziridinyl)phosphine oxide	<u>9</u>,*107*
Trypan blue	<u>8</u>,*267*

U

Uracil mustard	**9**,*235*
Urethane	**7**,*111*

V

Vinyl chloride	**7**,*291*
Vinyl chloride-vinyl acetate copolymers	**7**,*311*
4-Vinylcyclohexene	**11**,*277*
2,4-Xylidine and its hydrochloride	**16**,*367*
2,5-Xylidine and its hydrochloride	**16**,*377*
Yellow AB	**8**,*279*
Yellow OB	**8**,*287*
Zectran	**12**,*237*
Zineb	**12**,*245*
Ziram	**12**,*259*

www.ingramcontent.com/pod-product-compliance
Ingram Content Group UK Ltd.
Pitfield, Milton Keynes, MK11 3LW, UK
UKHW051259180426
11947UKWH00020B/1802